SCIENCE AND TECHNOLOGY IN AFRICAN HISTORY WITH CASE STUDIES FROM NIGERIA, SIERRA LEONE, ZIMBABWE, AND ZAMBIA

SCIENCE AND TECHNOLOGY IN AFRICAN HISTORY WITH CASE STUDIES FROM NIGERIA, SIERRA LEONE, ZIMBABWE, AND ZAMBIA

Edited by

Gloria Thomas-Emeagwali

The Edwin Mellen Press
Lewiston/Queenston/Lampeter

Library of Congress Cataloging-in-Publication Data

Science and technology in African history with case studies from
    Nigeria, Sierra Leone, Zimbabwe, and Zambia / Gloria Thomas
    -Emeagwali, editor.
        p.   cm.
    Articles chiefly in English; one article in French.
    Includes bibliographical references.
    ISBN 0-7734-9557-6
    1. Science--Africa--History--Case studies. 2. Technology--Africa-
    -History--Case studies. I. Thomas-Emeagwali, Gloria.
    Q127.A4S28   1992
    609.6--dc20   92-19201
                                                                       CIP

A CIP catalog record for this book is available
from The British Library.

The Edwin Mellen Press   The Edwin Mellen Press
P.O. Box 450   Box 67
Lewiston, NY 14092   Queenston, Ontario
USA   CANADA L0S 1L0

                    The Edwin Mellen Press, Ltd.
                      Lampeter, Dyfed, wales
                    UNITED KINGDOM SA48 7DY

            Printed in the United States of America

DEDICATED TO MY DEAR MOTHER

BRIDGET THOMAS

TABLE OF CONTENTS

Introduction .... *i*

Chapter One 1
Proverbs as Repositories of Traditional
Medical Practice in Nigeria
J. Olowo Ojoade

Chapter Two 23
Biologically-based Warfare in Pre-
Colonial Nigeria
Bala Achi

Chapter Three 33
Arithmetic in the Pre-Colonial Central Sudan
Ahmad Kani

Chapter Four 41
The Control of Water-Based Diseases
in Colonial Northern Nigeria
Gloria Thomas-Emeagwali

Chapter Five 55
Metallurgy in Northern Nigeria: Zamfara
Metal Industry in the 19th Century
Nurudeen Abubakar

Chapter Six 79
Gold Mining in Pre-Colonial Zimbabwe
L. R. Molomo

Chapter Seven 101
Diamond Mining in Sierra Leone 1930-1980
A. Zack-Williams

Chapter Eight 129
Glass making Technology in Nupeland, Central
Nigeria: Some Questions
Gloria Thomas-Emeagwali & A. A. Idrees

Chapter Nine 147
Science and Technology Policies in Zambia
Donald Chanda

vi

Chapter Ten 165
    Obstacles in the Development of Science
    and Technology in Contemporary Nigeria
        Julius Ihonbvere

Chapter Eleven 183
    Historical Perspectives on Technical Cooperation
    in Africa/Une Perspective Historique de la
    Cooperation Technique en Afrique
        Carlos Lopes

Contributors 203

LIST OF ILLUSTRATIONS

Euphorbia hirta.. 25

Strophantus hispidus.. 26

Bridelia ferruginea 29

Bronze plaque representing Oba Akengboi of
late 17th century Benin in battle dress 95

Cast Gold Alloy, Ivory Coast-
Late 17th century 96

Lady Adorned with Gold jewelry, Senegal
20th century   97

MAPS

Some Major town and villages in the Zamfara Area of
Sokoto, Nigeria 76

A Map of Nigeria, Ethnic Groups 77

A Map of Africa.. 78

ACKNOWLEDGEMENTS

This work is the product of African scholars from Nigeria, Sierra Leone, Guinea Bissau, Zimbabwe, and Zambia. I extend my special thanks to all of them for their faith in this project which would have been impossible without their co-operation. I also thank my numerous colleagues from Ahmadu Bello University, the Nigerian Defence Academy and the University of Ilorin, Nigeria, where I worked for a decade. I owe my intellectual inspiration to them no less than to my colleagues at St. Antony's College, Oxford where I served as a Visiting Scholar for a year, or to my present colleagues at Central Connecticut State University, New Britain here in the U.S.A. They have all contributed in one way or the other to the production of this text.

Gloria Thomas-Emeagwali
C.C.S.U., New Britain
1992

INTRODUCTION

In *A World of Propensities*, Karl Popper, the grand old sage of the Philosophy of Science, has pointed out that all organisms are problem finders and problem solvers and that knowledge is as old as life.[1] The present text is to a large extent an empirical representation of this proposition, albeit with reference to Nigeria and other parts of Africa. Logical consistency, the identification of regularity in nature and predictive capability are some of the variables associated with the scientific enterprise which has an a primary objective, the explanation of natural phenomena and physical reality. In the case of logical consistency we are concerned with the rules of reasoning as these are reflected in the understanding and explanation of the world. Science as a method of inquiry reflects organised thought, coherence and logical argumentation, and focuses on the unique and the particular only as this illuminates and facilitates the understanding of general trends and tendencies. Science is in constant search for uniformities and it is on the basis of this quest for generalities that scientific laws, whether written or unwritten constitute a major aspect of the scientific enterprise, making predictive capability a plausible attainment.

In this text we focus on some issues related to science and technology development in the region in a broad sense. Technology development is focused on in the context of metallurgy specifically not only in pre-colonial Nigeria but also in pre-colonial Zimbabwe and 20th-century Sierra Leone. Science and Technology policies in specific regions such as Zambia and

---

[1] Karl Popper, *A World of Propensities*, Thoemmes, Bristol, 1990, pp. 27-51.

Nigeria as well as Africa as a whole, constitute another area of research. The focus is primarily in terms of the development of material culture and the evolving constraints to continued technological growth in the regions chosen as case studies.

Two of the first three chapters are concerned with health issues in terms of historical and sociological perspectives primarily. We should note, however, that proverbs are themselves repositories of historical information insofar as they inform us of the modes of thought with respect to medical practice in the collective memory in pre-contemporary periods. There is greater time specificity in terms of the focus on the control of water-based diseases which is admittedly a pioneering work in the context of Nigerian historiography. Perhaps what is lacking most in this latter work is a more holistic perspective and interpretation in the context of the political economy of health but whilst recognising this weakness the present writer decided to leave it in its present form if only to stimulate other researchers into improving on the prevailing mode of interpretation and methodology. A third chapter focuses on what is referred to as biologically based warfare and here the emphasis is primarily on various techniques derived from trial-and-error experimentation and hypothesis formulation. The principle of immunization constituted an important aspect of indigenous science for the collective military forces, and so too the use of toxic substances in warfare.

There is major emphasis on the Nigerian region which engages as many as six chapters. One of the reasons for this is the editor's greater familiarity with Nigerian scholarship. It was therefore easier to mobilize responses with respect to specific issues. The response from Nigerian scholars may have been much less enthusiastic at the present time than two years ago when this was done, given the deepening debt crisis and the debilitating effects of the ongoing Structural Adjustment Programmes. There are, it seems, much less resources available now for research and writing not only in Nigeria but also in other parts of the continent, and yet there is a dire need for intensified research into a range of issues related to science and technology development in the region to facilitate a harnessing of indigenous resources and enhance the technological capacity of the continent. We need more in-depth studies into the development of mathematical thought, indigenous perspectives on time and space, traditional medical practice, and the reconstruction of indigenous techniques in textile, mining and engineering. In the final analysis it is the shrewd adaptation of some of these technologies that could lead to sustained economic growth.

Chapter I

Proverbs as Repositories of Traditional Medical Practice in Nigeria

J. Olowo Ojoade

AIM:

The intention in this chapter is to show how proverbs mirror traditional medical practice in Nigeria. Generally proverbs reflect the ideal or norm rather than the actual practice.

SOURCES:

For this study several collections of Nigerian proverbs were consulted. I have attempted to find in them proverbs that deal with medicine directly or indirectly. Out of the several examples I have chosen just a few for purposes of illustration.

VALUE OF STUDY:

The main value of this study lies in the fact that some raw data have been provided for both traditional medical practitioners as well as western medical practitioners including medical anthropologists. For that matter there are also scholars working on various aspects of African culture. They will also, I believe, benefit from the study.

The proverbs herein listed shed light primarily on traditional medical practices and focus also on other aspects of Nigerian lifestyle. Indeed nowhere else better than in the current sayings of everyday life can the attitude of any given people towards a particular subject be observed.

That being so, the value of the study of proverbs dealing with African traditional medicine needs no defence.

It is perhaps needless to state that the typical African has greater belief and confidence in the ancestral experience which emanates from centuries of practice and trial and error than in the medicine brought by the white man. A western doctor treating such a person must therefore bear this psychological point in mind. And as the epigraph clearly indicates the marriage of western and traditional medicine is the greatest and most urgent desideratum now for the benefit of mankind.

APPROACH:

A popular approach among serious folklorists is to print the proverbs in their original texts. But space and printing limitations indeed preclude this step. I have done so only in respect of Yoruba proverbs.

THEME I: THE TRADITIONAL MEDICAL PRACTITIONER.

There are different types of traditional doctors. For example, the Yoruba have two main types: the Ifa priest called babalawo, whose speciality is divination and psychotherapy, and the herbalist otherwise called onishegun. According to Maclean (1971:75):

> On the whole it would be true to say that the onishegun are less highly regarded than the babalawo, the former dealing with the commoner, more easily recognized disorders whilst the latter take on more of the difficult diagnostic and prescribe more radical regimens for a patient to follow.

The first two proverbs allude to the doctor's personality as well as his knowledge of drugs.

> Even if the doctor does not have charms his eyes are more frightening than charms. (Idoma)

> The sick man knows the doctor is the one who knows the root for the drugs. (Idoma)

One proverb refers to young medical practitioners:

> A young person should practice medicine in secret. (Edo)

Because it is normally old men who are experienced enough to practice medicine, younger ones have to be apprenticed. Indeed as Maclean (1971:76) noted, all have to undergo "a prolonged period of apprenticeship or training before feeling qualified to set up practice on their own."

The doctor must of necessity obey some rules. Indeed there are associations of traditional doctors who make strict rules which are strictly followed. Some of such rules (*cf.* Simpson, 1980:95) include:

(i)   The doctor must not use poison.

(ii)   The doctor must not tell lies such as that doctors can make money by means of medicine.

(iii) The doctor must not collaborate with thieves.

(iv)   If a woman is brought to the doctor, he should not have sexual affairs with the woman or marry her, but if he wishes to marry her he must get a promissory note from her family.

(v)   If a pregnant woman comes to consult the doctor, he must not use bad medicine to procure an abortion for her.

With the above must be compared the Hippocratic Oath of which these rules are strikingly reminiscent.

> When a doctor prepares a concoction he tastes it (first). (Ukwani)

Most times he is successful though it is believed that destiny has to do with whether the patient survives or dies.

> The doctor said it was his doing (but in reality) (it is because) death refused (to take the patient). (Idoma)

The proverbs also manifest the doctor's reactions when he has failed:

> The doctor is in sorrow as the patient is about to die. (Edo)

That is, one source of his income will be lost.

> If a doctor has made a mistake he leaves by the back of the house. (Igbo)

That, sometimes, the doctor cannot cure himself is clear from the subjoined proverbs:

> When a doctor is sick it is a layman that gives him medical treatment. (Edo)
>
> Ibi olosigun ba nogbon de, oun r'o ma fi jerun mo. (Ijebu/Yoruba)
>
> It is the vastness of a doctor's wisdom that determines his ability to satisfy his hunger for food.
>
> The native doctor is a kinsman to spirits. (Birom)
>
> After god comes the native doctor. (Ndoni)

THEME II: FEES.

Doctor's charges which are paid either in cash or in kind or both are also attested by proverbs. Only one proverb specifically alludes to a fowl or goat as one of the things given to the doctor for a cure. Other proverbs allude to fees charged by the doctor in a general way. The following are some of the pertinent proverbs:

> When a medicine man is well fed (i.e. well paid) he collects herbs from the thickest forest. (Igbo)
>
> The wealth of the sick man is the doctor's. (Fulani)
>
> A sick man is a slave to the doctor. (Ijo)
>
> Without paying the doctor's fees sickness will not end. (Edo)
>
> Doctor, whoever dies has underpaid you. (Hausa)

That is, pence are not counted in time of illness.

> Expensive medicine sticks to the gullet (when you remember what you had paid for it). (Hausa)
>
> You ruin your house if you listen to all that the doctor says. (Ukwani)

Because the doctor at times asks for too much:

Any one who moves with the native doctor always gets the leg of a fowl. (Urhobo)

To be a doctor is to know how to eat chicken. (Jukun)

The more goats you give to the doctor, the better your treatment. (Yeskwa)

You promise a goat when you wish to be well, but a fowl will do once you are well. (Hausa)

THEME III: TREATMENT.

Some proverbs allude to how treatment is given to patients. The following proverbs are some of the many that indicate the method of treatment.

Sa' a bi ologun ti wi. (Yoruba)

Do exactly as the doctor has prescribed it.

If you intend to give a sick man medicine, let him get very ill first, so they he may see the benefit of your medicine. (Nupe)

The doctor says medicine needs endurances. (Edoi)

Oogun ki igbe inu ado je. (Yoruba)

A medicine cannot stay inside its container and be effective.

Oogun ta a ko fi owo se, ehin aaro ni i gbe. (Yoruba)

The medicine that is prepared without money (being spent) stays behind the hearth.

That is such medicine will not have much value.

Gba mu ko tan iba. (Yoruba)

'Take some and drink' does not cure a fever.

That is, don't ask for a portion of medicine from another patient. Rather go yourself and see the doctor for yours.

We apply the medicine before the sickness arrives. (Jabo)

Everyone knows that fresh tribal marks should be treated with hot water; but it should not be too hot. (Edo)

The proverb means that there is no pleasant way of effecting cures. To help a person with a fractured limb, the surgeons have to amputate at times. A person suffering from appendicitis is cut open in order to save him. A very bitter mixture may be prescribed for a certain ailment. To remove a thorn from the foot, more pain must be inflicted in the process. Thus when a patient complains that the medicine is too bitter, that the treatment is too harsh, etc., he is promptly reminded that one must suffer in order to be relieved of suffering. (*cf.* Ojoade 1970:26)

> 'To say "this is medicine for a headache" is to know that it does cure.' (Sena)

The meaning is that the native herbalist must be given time to go and search for the suitable plant or root.

> Oojo ti a ba nwo alaisan ni a nwo ara eni. (Yoruba)
>
> It is at the time the doctor gives treatment to the patient that he himself treats himself.
>
> It was a frog that was taken wherewith to test the arrow poison. (Idoma)
>
> Relapse is not good for a patient. (Edo)
>
> When a doctor does not wish to continue with a sick man or collect more money he says that the cause of the sickness is at the bottom of the sea. (Ndoki)
>
> If the medicine can't be drunk it can be poured away. (Hausa)
>
> Ki a to je ogun, ki a to bi i, ono ofun a ti be. (Yoruba)
>
> By the time you swallow medicine and then vomit it, the throat is already bruised.

That is, it is better not to swallow it at all.

THEME IV: INGREDIENTS.

Allusions are also made to some ingredients which the doctor uses for his concoctions. Some works have been done on and collections made of African medicinal plants. Notable among the collectors are Dalziel (1937), Githens (1948), and Sofowora (1979). Pierre Verger's (1967) stupendous work contains 3,000 Yoruba plant names and formulae, although he still believes that he has not exhausted the list. Says Beier (1970:122):

> The Yoruba have a special orisha for medicine, called Osanyin. The priests of Osanyin, and likewise the Oracle priests, must have a thorough knowledge of the leaves and fruit of the forest

and their medicinal uses. The Yorubas have known many potent medicines, poisons and sedatives for a long time; but their use of plants is a mixture of "scientific" medicine in the Western sense and sympathetic magic. The use of a plant must be accompanied by the appropriate magic formula.

Subjoined are some illustrative proverbs:

When a medicine man is well fed, he collects herbs from the thickest forest. (Igbo)

He who is in need plucks leaves from the bush. (Ijo)

I.e.: It is a sick person that approaches a doctor.

The doctor does not tolerate the unnecessary burning of the bush,

I.e.: Because he picks his leaves from that bush.

Pepper should not be added to a medicine meant to cool the eyes. (Igbo)

Ara logun emi, emi ti ko je ata, emi yepere. (Yoruba)

Pepper is the medicine for health, the person that does not eat (take) pepper is useless.

Where the medicine man lives alligator pepper is always found there. (Igede)

I.e.: Because alligator pepper constitutes an important ingredient in the doctor's medicine.

Atare ri eni tun idi re se, o nfi oburo sesin, oburo iba ri eni tun idi re se, a sunwon ju stare lo. (Yoruba)

The alligator-pepper has someone to take care of it, and it is mocking the oburo tree; if oburo tree had someone to tend it like the alligator-pepper tree, it would be more valuable than the alligator-pepper tree.

(The Yoruba farmer tends the alligator-pepper tree because its seeds are used for medicinal purposes.)

There is always kola-nut in the medicineman's bag. (Igbo)

Iyan ni onje, oka li ogun, airi rara l'a nje eko. (Yoruba)

Pounded yam is real food, oka is as good a food as a medicine, it is when there is no other food at all that we eat eko.

(Eba or oka is a mixture of powdered cassava (gari) and hot water; eko is cooked corn starch.)

Medicine that is mixed with food, even if it does not cure the disease, will cure hunger. (Nupe)

It is with old honey that medicine is made. (Hausa)

A native doctor asks for a human head when the prospect of survival of a sick person is remote. (Ukwani)

A native doctor never forgets his bell. (Okrika)

(Used to summon the spirits?)

THEME V: EFFICACY OF MEDICINE.

Yes, medicine is useful and efficacious. But African proverb-coiners also believe that medicine need not be employed all the time. The user need not be a slave to it. For example they say:

Sari ma sagun, ogun li ojo iponju, ori li ojo gbogbo. (Yoruba)
Use your head and not medicine. Medicine is for the time of trouble but the head is for all time.

and

Inun mimo ju oogun lo. (Yoruba)
A clear mind is more efficacious than mere medicine.

Ori ja ju oogun. (Yoruba)
One's genius fights better than medicine.

THEME VI: PATIENTS.

The patient and his idiosyncracies form the subject matter of the next set of proverbs.

Alaisan l'nwa oloogun (Yoruba)
It is the person who is ill that looks for the doctor.

It is a sick person that approaches a doctor. (Ijo)

Onen ara re ya re ma ubo aisan ti mi se ghun rin. (Ilaje)
It is the person who is afflicted by a disease that knows what part of him is affected and how the disease affects him.

Onen aisan mi se, jo fi pama, ara re ghon mi se rin. (Ilaje)
A person who is attacked by a disease and hides it, it is himself he is undoing.

A patient does not term his ailment sickness, he says: "just only this that pains me." (Edo)

E ma aisan otosi. (Ilaje)
One does not (easily) know when a pauper is ill.

Because, even though he is ill he will still have to fetch food for himself. Today, for example, he says, 'I have fever.' Yet he gets up to cook his food. Afterwards he goes near the fire to warm himself. At another time he says he has cold. Yet he has to prepare his own food. So he goes to the river to catch fish. When he returns home, he goes near the fire again and so on and so forth.

Alaisan ti o ba gbon ju ologun a soro iwo. (Yoruba)

The patient that is cleverer than his doctor is difficult to cure.

Bi o ti wu ki o ri, a ki rerin abiron; boya chun ti a se e loni a se iwo lola. (Yoruba)

One should never laugh at a sick person; perhaps what afflicts him today may afflict thee tomorrow.

A mbe ologun a ko iti be eniti o ma a muu. (Yoruba)

One is begging a doctor to concoct medicine for a sick person (but) one has not been assured that the sick person will accept to take the medicine.

Ise re ni, ko je ki a mo iku abiku. (Yoruba)

Its general behavior prevents us realizing the death of an abiku child.

(An abiku child usually faints when it is ill. When it actually dies people may think it has only fainted and that it will come round again).

Bi olokunrun yio ba ku, ki o ma puro mo alapa, omitoro ki ikoro. (Yoruba)

If the sick man is going to die, he should not tell lies against the soup; no gravy ever tastes bitter.

Enu alaison l'o fi npe iku. (Yoruba)

The sick person invites death with his own mouth.

(That is, by losing hope and sometimes, for example, when he coughs, he might exclaim: Oh, God, this cough will kill me!)

Some who are desperately ill may even wish that they had rather die than live!

A sick man blows a dog to death with his fist and he complains that ill-health has defamed him.

Suppose he were not ill he would have destroyed the king's subjects. (Edo)

Ara iya o ya Ara iya a ya. Edo nwole, Ewe njade. (Yoruba)

Mother is not well. Mother is (indeed) not well. (But Eko is brought into her room and it is only leaves that leave the room.)

This is used to ridicule a sick person who eats more than a healthy person. Eko is similar to blanc-mange and is usually wrapped in leaves.

Bi o ti wu ki o ri, a ki irerin abirun; boya ohun ti a se e Iloni le se 'ni l'ola. (Yoruba)

Whatever may happen, it is not proper to laugh at a deformed person; perhaps what afflicts him today may afflict us tomorrow.

If you have never been sick, you boast of your strength. (Jabo)

When a sick patient gets a relief he curses the king. (Edo)

Ara tu olukunron o gbagbe opa. (Ilaje)

The sick man gets well, he forgets his walking stick.

THEME VII: HEALTH.

Health is what the traditional practitioner wishes to maintain. So what do the proverbs say about it?

Ara lile l'ogun oro. (Yoruba)

Health is wealth.

Abarapara ti gbogbo enia, okunrun ti eni nikonsoso . (Yoruba)

The healthy are much in demand, but the sick are neglected.

THEME VIII: PREVENTION.

The idea that "prevention is better than cure" must date back to the remotest antiquity. Africans express this principle in many ways including their medical proverbs.

A ki ifi itiju k'arun. (Yoruba)

Don't let politeness make you run the risk of contracting disease.

Arun e ju pa ma'ra. (Ilaje)

One does not keep secret what one is suffering from.

(Otherwise one may die.)

Bi ara ile eni nje kokoro buruku it a ko ba tete wi fun u, herehuru re ko ni jeki a sun l'oru. (Yoruba)

If a kinsman is not warned in time when he eats poisonous insects, the resulting itch will keep the whole family awake at night.

Bi ako li aiya rindo rindo, aki ije ayan. (Yoruba)

If the stomach be not strong, do not eat cockroaches.

Elesin se ogun ejo tan, elesin wa l'oke, ejo wa nisale. (Yoruba)

A man on horseback has taken effective measure to avoid a snake bite; he is on the back of the horse, the snake is on the ground.

Ifura l'ogun ogba. (Yoruba)

Caution is the medicine of an elder.

THEME IX: DISEASES (ILLNESSES).

The concern of the next set of proverbs is about diseases, illnesses or sicknesses, especially the serious effect of them. Thus:

Arun l'a a wo, a ki iwo iku. (Yoruba)

It is sickness that can be cured, not death.

A e ri kekere aisan. (Ilaje)

No illness is small. Therefore one should not despise an apparently small illness. If not treated in time it may develop into a more serious illness.

Arun mefa ki i ba agba ti: ko je, ko mu; ko su, ko to; ko sun, ko wo! (Yoruba)

Six different diseases do not attack an elder: he cannot eat, he cannot drink; he cannot defecate, he cannot urinate; he cannot sleep nor can he keep awake.

Arun t'o ns'ogoji l'o ns'odunrun, ohun t'o nse Aboyade, gbogbo Oloya l'o nse.

A disease that affects forty also affects three hundred. What befalls Aboyade (a child given in answer to prayer to the god of river Niger) also befalls all the worshippers of Oya.

This is used in the same sense of "what is good for the geese is also good for the gander."

He who does not sleep with sickness does not know how bad sickness is. (Ijo)

A rare disease is not cured with common herbs. (Igbo)

Disease will wipe out a town sooner than war. (Jabo)

If the Gba sickness attacks one, he does not (even) weep.
(Jabo)

(Because it is just too serious.)

E si ejo oneku, aisan re poju rin. (Ilaje)

It is not the fault of the dead person; it is the illness that is too much for him.

The remedy for a pestilence is plenitude of life. (Hausa)

THEME X: VENEREAL DISEASES.

Two venereal diseases are referred to by proverbs in my collection - gonorrhea and syphilis:

Atosi aisan gbajuma. (Ilaje)

Gonorrhea, the disease of a gentleman.

Atosi afa amutorunwa ni. (Yoruba)

Gonorrhea caught by an Imam was brought from heaven.

This to ridicule Muslim preachers who are supposed to be chaste.

Atosi e e duro n'ara omoten (Ilaje)

Gonorrhea does not stay in the body of an alcoholic.

It is strongly believed that a constant drinker of alcohol hardly catches gonorrhea.

The person suffering from syphilis knows how to sleep with his wife. (Idoma)

THEME XI: FEVER.

Fever, particularly malaria fever, is also alluded to:

Eniti iba nse ki i je epo, oniko ki ije obi. (Yoruba)

A person suffering from fever does not take oil, the one coughing does not eat kola.

This needs no explanation.

THEME XII: LEPROSY.

Leprosy, a terrible disease, perhaps the disease carrying the greatest stigmata (Foster and Anderson, 1978:160), is also alluded to by the proverbs. It has been the practice in Biblical times and even in most modern societies for lepers to stay outside the city. Traditional Africans also made lepers live outside the community - away in the bush.

> Amu ni se esin; ete ti imu ni li agogo imu. (Yoruba)
>
> The slanderer brings disgrace to one, like a leprosy which attacks one on the point of the nose (where all can see it).
>
> Oghun ti o jo oghun ni a fi nwe oghun ni ko je ki a fe ete we lapalapa. (Yoruba)
>
> Whereas we can compare two things that are alike, we cannot compare leprosy with ringworm (because they are not alike in that one is more serious than the other).
>
> Aki ifi ete sile ma pa lapalapa. (Yoruba)
>
> One should not neglect a patient's leprosy and begin to treat his ringworm.
>
> "So we began" says a leper when he sees a man with ringworm. (Hausa)
>
> Eniti o ba adete gbe'le oun ni enitiowo ndun. (Yoruba)
>
> It is the person that lives in the same house with the leper that will have the disease in his hand.
>
> The leper isn't a stranger to sores. (Hausa)

The proverb is used in the sense of "don't teach your mother to suck eggs."

> A ko ni ki omode ma ya adete ti o ba ti le da gbe ninu igbo. (Yoruba)
>
> We do not stop a child from having leprosy provided he can stay alone in the forest.

THEME XIII: INCURABLE DISEASES.

African traditional doctors have, with the passage of time, discovered, like their western counterparts, that certain diseases are either very difficult to cure or are absolutely incurable. The following proverbs make allusions to such diseases:

Asoro wo bi arun idile. (Yoruba)

Difficult to cure like a hereditary disease.

Aisan ti ko gbo oogun ni nmu ni mo egberun awo. (Yoruba)

It is an incurable disease that makes the patient consult several doctors.

Arun ti ko se iwo, Olorun ma fi se ni. (Yoruba)

A disease that defies cure, may God not allow it to attack us.

A disease that has never been seen before cannot be cured with everyday herbs. (Igbo)

Ogbo ko l'ogun. (Yoruba)

Old age has no medical remedy.

THEME XIV: MADNESS.

Among the principal causes of madness according to the Yoruba are punishment from gods or spirits because the victim has broken some taboos or failed to perform certain obligations like ceremonies. Madness can also be caused by witchcraft; it can be inherited by the victim from his parents (Leighton, 1969:184-185).

The proverbs in this section dwell on the cure of madness. Can it be cured completely? Traditional medicine believes that a madman never returns to full normalcy:

A lunatic is harder to find a cure for than a madman. (Hausa)

That is, the hardest madman to deal with is the man not mad enough to put in an asylum.

If the madman stops his madness he does not stop soliloquies. (Igbo)

Madness can be cured, but the murmuring aspect of it cannot be cured. (Ijaw)

A madman cannot be cured; (he) only (gets) a little better. (Hausa)

THEME XV: DIVINATION.

Divination is a sine qua non in the healing process employed by the Yoruba folk doctor. According to Professor Idowu (1963:77-78):

> Before a betrothal, before a marriage, before a child is born, at the birth, at successive stages in a man's life, before a king is appointed, before a chief is made, before anyone is appointed to a civic office, before a journey is made, in time of crisis, in times of sickness, at any and all times, Ifa is consulted for guidance and assurance.

One would have been surprised if this phenomenon was not manifested in proverbs. The objective behind divining is to discover, for example, the cause(s) of an illness, what sacrifices to offer to the god(s) that causes or cause the illness, etc., and what herbs to use in the healing process.

The following are some relevant proverbs:

> Owe ni Ifa npa, omoran ni mno o. (Yoruba)
>
> Ifa speaks in parables; it is a wise person that understands the speech.
>
> Bi oni ti ri ola ki i ri be ni babal wo fi ndifa Ororun. (Yoruba)
>
> Because as today appears tomorrow will not appear; that is the reason why an Ifa priest consults his oracle every five days.

This is to enable him to know what may likely happen in the future.

> Ojumo ki imo ki Awo ma sode wo; agbede a gbon adal. (Yoruba)
>
> Morning does not break but the Ifa priest consults his oracle; similarly morning does not break but the blacksmith is requested to sharpen matches for farmers.

THEME XVI: SACRIFICE.

In many traditional African societies, sacrificing constitutes a major step in the healing process. Divination is resorted to in determining the cause, the ingredients and the mode of cure of many diseases. Therefore, usually sacrifices form an important stage in the healing process.

Indeed even western-trained psychiatrists like the Nigerian Professor T. A. Lambo, formerly Deputy Director General of the World Health Organization, who have adopted the "unorthodox collaboration with the traditional healers" and have for a number of years made use of the services of the so-called 'witch doctor', have confirmed that often, especially in cases of psychotherapy with those who have psychoneuroses and anxiety states, rituals and sacrifices "have sufficed." (Lambo 1964:447ff.).

Below are a few proverbs relating to sacrifice:

A ki idifa k'oma yan ebo. (Yoruba)

The diviner does not consult Ifa without prescribing a sacrifice.

Ebo die, Ogun die ningba alaiku la. (Yoruba)

A little sacrifice, a little medicine saves the patient from death.

Babalawo ki bere ebo ana. (Yoruba)

The doctor does not ask for the sacrifice made the previous day.

That is, another will (surely) be made today.

Obuko ni aisan agbe lowo on yi mba on leur; bi aisan na ba po si babalawo a ni ki nwon lo mu obuko wa lati fi se etutu fun u; bi o ba san, awon omo re a ni nwon o mu obuko fi wewu amodi. (Yoruba)

The he-goat says that the sickness of the farmer, his owner, is making him feel very frightened. If the sickness gets worse, the native doctor will say that a he-goat should be brought in order to make propitiation for him; if he gets better, his children will say that a he-goat be brought in order to make thanksgiving for the escape from illness.

He-goat, rather than the female is the usual sacrificial animal among the Yoruba.

Effective drug does not require much sacrifice. (Igbo)

Let us continue to sacrifice to the gods so that they become ashamed. (Igbo)

Like the Ancient Greeks or Romans traditional Africans also assign certain functions to different gods. For example, we have the god of war, god of the seas and rivers, god in charge of food etc. The above proverb, however, is referring to the god in charge of childbirth particularly. A barren couple was advised to make sacrifices to this god who, on the principle of quid pro quo, is expected to give them a child. If after so many sacrifices there was still no pregnancy, this proverb would find application. The couple had done their part of the contract; the god (or gods) had broken the contract, hence the god(s) would become ashamed.

THEME XVII: DESTINY.

The part played by destiny in ensuring a cure for the patient or in ensuring death for the patient is clearly indicated by the proverbs. Thus:

Iku ti ko ni ipa ni ni igbe alawo rere ko ni. (Yoruba)

When death is not ready to receive a man, it allows an expert physician to come at the right time.

On the day of death, the useful medicine is never available. (Tiv)

Or

Medicine will not revive him doomed to die. (Hausa)

Or

Bi o ni opo ogun, bi o l'eke ko ni je; ori eni je ju ewe lo; ipin ja, o ju ogun lo. (Yoruba)

If you have magic charms and you are a gossip, the charms will not be effective; a person's destiny is more powerful than charms; what will be one's station in life cannot be altered by charms.

THEME XVIII: DEATH AND SICKNESS.

It is appropriate to bring the question of death, the end of all living organisms, last.

Where sickness abides, there death abides. (Jabo)

Eni ti aigbon pa l'o po, eni ogon pa ko to nkan. (Yoruba)

Those killed by lack of wisdom are numerous. Those killed by wisdom do not amount to anything.

Atete-sun ni ateteji, eni tete ku lo mu 'le lorun. (Yoruba)

Early to bed is early to rise. One who dies young goes slightly sooner to take a plot of land in heaven.

Osan orun ko pon eni ti o ba setan ki o ma lo. (Yoruba)

Though the day for going to the other world is still young, he who is ready may start on his way. It is never too late to die, one who is ready may die now if he chooses.

Aide iku ni a nso aja morun; bi iku ba de, aja a sonu; a gbe alaja lo raurau. (Yoruba)

It is when death has not come to a patient that aja on his neck is effective, but when he is actually due to die, the aja will become the life of the patient even with aja on his neck.

(Aja is juju to prevent death).

Death does not reason with its victim. (Igbo)

Iku ko gbo oogun. (Yoruba)

When a patient is going to die there is no need for medicine.

With this we must compare the Latin "Contra vim mortis non est medicamen in hortis."

The above is an example of a medical proverb that is international in its currency.

THEME XVIIII: DEATH AND THE DOCTOR.

By the time the doctor's child has died so much medicine (leaves) must have been brought from the bush into the house. (Tiv)

The native doctor cures one disease but another disease kills the native doctor. (Igbo)

Medicine men and their clients are eventually victims of death. (Igbo)

Alawo hio ku, onisegun yio re orun, adahunse ko ni gbele. (Yoruba)

The witch-doctor will die, the medical doctor will go to heaven and the consultant will not be left behind.

Iku pa babalawo b'eni ti ko n'ifa. (Yoruba)

Death kills the medicine man like someone who is completely ignorant about the art of divination.

CONCLUSION

In conclusion I wish to stress the same points which I had stressed in my paper on the Madman (Ojoade, 1983) and other works (Ojoade 1980a, 1980b, 1980c, 1981, 1982, 1983, 1984), namely that proverbs and proverbial sayings of all cultures, not only those of Africa, be collected and scientifically annotated. For in them we see what peoples of antiquity had done and how they had done them. In short proverbs constitute the macrocosm of the culture that uses them. The fact that, in spite of the introduction of Western medicine, people still insist in using traditional medicine, amply shows that they believe in this heirloom.

I suggest that more research should be done on them to ascertain what truths inhere in them. But it must be said at once that a doctor who has been prescribing a particular drug for years must have certified its efficacy, otherwise he would have discarded it. Similarly a traditional medical concoction which is still being prescribed must have stood the test of time in its efficacy. It is with this fact at the back of my mind that I strongly suggest that both traditional and western medicine be utilized complementarily. The necessity of knowing what the people believe in and so practice religiously is of paramount importance.

It has now been confirmed that "western psychiatric techniques are not...demonstrably superior to any indigenous...practices" (Prince, 1964:116). It is also true that patients unsuccessfully treated with western methods have obtained complete cure from traditional healers.

20

REFERENCES

Ajibola, J. O. *Owe Yoruba*. Ibadan O. U. P. 1976.

Beier, U. *Yoruba Poetry*, Cambridge University Press, 1970.

Champion, S. G. *Racial Proverbs*. London. 1950.

Dalziel, J. M. *The Useful Plants of West Africa*. London, The Crown Agents for the Colonies. 1937.

Foster, G. M. and Anderson, B. G. *Medical Anthropology*. New York. 1978.

Githens, Thomas B. *Drug Plants of Africa*. Philadelphia, University of Pennsylvania Press. 1948.

Idowu, E. B. *Olodumare: God in Yoruba Belief*. New York. 1963.

Kiev, Ari (ed.). *Magic, Faith and Healing*. New York, The Free Press of Glencoe. 1964.

Lambo, T. A. "Patterns of Psychiatric Care In Developing African Countries." In *Kiev*. 1964.

Leighton, A. H., "A Comparative Study of Psychiatric Disorder in Nigeria and Rural North America." In Plog S. C. and Edgerton R. B. 1969.

Maclean, Una. *Magical Medicine: A Nigerian Case-Study*.

Harmondsworth, Middlesex, Penguin Books. 1971.

Nyembezi, C. L. S. *Zulu Proverbs*. Johannesburg. 1974.

Ojoade, J. O. "Observations on Medical Practice in Roman Africa." In *Nigeria and the Classics* Vol 12: 19-30. 1970.

Ojoade, J. O. "Some Ilaje Wellerisms." In *Folklore* (British) Vol. 91 No. 1: 63-71. 1980a.

Ojoade, J. O. "Some Itsekiri Proverbs." In *Nigerian Field*. Vol. 45 pt 2/3 (July): 91-96. 1980.

Ojoade J. O. "African Proverbial Names." In *Names* Vol. 28 No. 3 (Sept.): 195-214. 1980c.

Ojoade, J. O. "Nigerian Animal Proverbs and Their Importance. The Tortoise as depicted in the Life of Ilaje Africans." In *Folklore* Vol. 33. No. 2: (February. First Part) 37-44. 1982.

Ojoade, J. O. "African Sexual Proverbs: Some Yoruba Examples." In *Folklore* Vol. 94 No. 2: 201-213. 1983.

Ojoade, J. O. "Specimens of African Humor and Wit." Paper read at the 1984 Western Humor and Irony Membership Conference on Contemporary Humor, held at Ramada Inn, 3801 East Van Buren, Phoenix Arizona 85008, U. S. A. 29th March, 1984.

Plog, S. C. and Edgerton, R. B. *Changing Perspectives in Mental Illness.* New York. 1969.

Prince, R. "Indigenous Yoruba Psyschiatry" in Ari Kiev (ed.), *Magic Faith and Healing.* New York, The Free Press of Glencoe, 1964.

Simpson, George E. *Yoruba Religion and Medicine in Ibadan.* Ibadan University Press. 1980.

Sofowora, Abayomi. *African Medicinal Plants.* University of Ife Press. 1979.

Taylor, A. *The Proverb.* Hatboro. 1962.

Taylor, W. E. *African Aphorisms.* London. 1924.

Verger, Pierre. *Awon Ewe Osanyin (Yoruba Medicinal Leaves).* Ife, Institute of African Studies. 1967.

Whitting, C. E. J. *Hausa and Fulani Proverbs.* Lagos. 1940.

Chapter II

Biologically Based Warfare in Precolonial Nigeria

Bala Achi

Biologically based warfare is an age-old practice in the Nigerian area. New murder weapons were introduced in warfare to give the assailant the advantage. These lethal biologicals gave their possessors both offensive and defensive capability and forced societies to devote more time and resources to military research leading to development in this sphere. Some of these biologicals were not basically lethal but produced temporary and reversible incapacitation such as temporary mental confusion and paralysis. Their impact depended on individual variables like age, body weight, sex and state of health. For example, infants and the infirm suffered more from them than others. The biologicals affected group cooperation, so that war leaders on whom these were applied, were forced to make illogical, and incoherent decisions and give irrational orders. This forced some communities to keep up with the scientific and technological progress of other communities. The use of biologicals in the warfare of precolonial Nigeria is a reflection of the scientific and technological development of society. The development of arms, armour, erection of static fortifications, the development of new tactics and strategies and the general organization of society in anticipation of danger, led to the employment of biologicals as a counter measure.[1] It was therefore the rate and direction of scientific and technological development that influenced tactics and dictated research in this field. The same environment that provided the resources for the technological development of society equally provided the animals and plants whose toxins were exploited for the preparation of biologicals. People realized that animals inject poison from their bodies into their enemies. Bees, wasps and ants sting and by so doing inject poison into the body. Spiders and snakes do the same with their fangs. For example, through the fang, the cobra emits haemotropic poison which acts on crythrocytes and destroys the red blood corpuscles of the victim. Toads give off a milky fluid which can make the victim sick or even kill him. These were effectively exploited in the offensive and defensive plan. In order to extract poison from these insects and reptiles, their heads and tails were cut off, crushed and dried. The content was then mixed with water and smeared on arrows and spears. Psychic forces believed to be impregnated in plants and mineral substances were equally consulted.

Arrow poison is one of the earliest biologicals to be used. This was a concentration of toxin from various leaves, barks and roots as well as heads of poisonous snakes millipedes and caterpillars.[2] This concentrated poison was known as 'dafin zabgai mai kare dangi' in Hausaland. (Poison that can destroy a whole generation.) It was so named because one arrow smeared with this poison could kill many people as long as the poison was introduced into the blood. When the victim was struck and a wound inflicted, some quick-acting poison would be introduced. This would make the victim lame and cause paralysis of the heart. Arrow poison was apparently more effective when arrows were tipped with iron. The barbs aided laceration while the longer shaft made penetration more effective. Iron-tipped arrows were relatively heavy and could hardly be deflected by wind. They could therefore hit a moving target at four hundred feet away. The addition of poison on these iron-tipped arrows made them a powerful offensive weapon. The earlier stone and wooden missiles were, however, hardly smeared with poison. Horses used in war by the enemy and the latter's water supply were also poisoned by biological agents.[3] Even though biologicals needed an incubation period before reaction could begin and could not be distributed over a wide area, they were effective in warfare.

Biologicals were not only used to poison the opponent, they were also employed to embolden the user and make him fearless in battle. This was necessary because fear has destructive effects. Fear could lead to psychiatric combat casualties like trembling and excessive perspiration among the fighting men. Such reaction inhibited warriors from bringing out their heroic attributes toward the war effort thereby affecting performance and survival of the military group. Thus, precolonial warfare was not just an intellectual art where people seized the advantage of weather and topography, but also scientific, requiring the employment of up-to-date weapons and biologicals. People were encouraged to become conversant with the medicinal uses for plants, in order to use such botanical knowledge for the continued survival of society. It indicates that no limit was ever reached in the search for potent snake venom and esoteric plant toxins. Some of the plant toxins exploited and used in the warfare of precolonial Nigeria included *Euphorbia hirta, strophantus hispidus, Bridelia ferruguinea, Calabar beans, Erythrophlem guinense and Dichrostachy glomerata.* An examination of some of these will reveal their toxicity and effects on war victims.

Euphorbia hirta (Nonon Kurciya.)...Used as arrow poison. It causes asphyxia and death.

Euphorbia hirta has stout stems with bristly hairs and prickly growths. It has a latex-like opium got by incision of the stem. It is emetic and powerfully cathartic, though used externally to act as a vesicant. Thus, while it is used as poison, it is equally used as an antidote to arrow poison and to treat asthma. It contains resin, calcium oxalate crystals, sugars, mucilage, oil, ceryl alcohol as well as linoleic acids and alkaloid. Thus, it has a depressant action on the heart and respiration while causing relaxation of the bronchi by central action. At high doses, it causes asphyxia, severe inflammation of the alimentary canal and vomiting.[4]

Strophantus hispidus produces pseudostrophantrin which causes vomiting and death. The seeds of the yellow fruits contain poison.

Strophantus hispidus is a genus of woody climbers with twisted petal tails and conspicuous flowers in wet and dry environments. It has long, white hairs with few spikelets. It yields pseudostrophanthin that is extremely toxic. It also contains glycoside mixture, K-strophanthin, alkaloid trigonelline, fatty oil, resin and enzyme. The dried ripe seeds contain 10% of a mixture of glycosides with 25% of fixed oil. It is these that have a terrible effect on the heart as it forces the heart to work faster. When used as poison, it causes paralysis of the heart and the body becomes black and sullen with blood issuing from the nose, ear and mouth. Within thirty minutes, the victim would die.[5]

Erythrophlem guinense is a leafless acuminate with oblong fruits. It is woody and leathery. The bark is sassy and contains a poisonous alkaloid and erythropheloeine. Calabar bean or ordeal bean contains alkaloid physostigmine, eserine, eseramine, geneserine, cabatine and calabacine. The chief alkaloid physostigmine, is present to an extent of 0.15 percent, derived from tryptophan. It has a myopic effect on the victim making him incapable of fighting. Biologicals used in the warfare of precolonial Nigeria were thus, of high strength. They could kill a normal person between a half and one hour after being struck by a missile smeared with such as long as the poison was in contact with the blood. Handling the poison could, however, be safe if there were no open wounds. Drinking could cause mild stomach disorders.

Biologically based charms were also used in precolonial warfare. Some of these were believed to make the wearer invulnerable to the weapons of the opponent. Others were believed to embolden the wearer and at the same time cause terrible fear and panic. There were charms that were believed to make ineffective the weapons of the enemy or cause weapons to melt, refuse to fire or break. It has been claimed that some charms could make the wearer become invisible if suddenly surrounded by the enemy.[6] These charms were often written in the Arabic script and wrapped in cotton. Charm dresses were usually sleeveless with a rectangular upper part and circular neck hole.[7] Even horses were provided with biologically based charms.

Closely related to the biologicals were egogenic aids used by societies in the Nigerian area in times of war. These were different types of plants whose roots and barks were soaked in water for days to allow for fermentation. The mixture was boiled with corn into porridge to help improve conditioning and performance. This concoction was believed to be capable of enhancing the ability of warriors to sustain high intensity combat. The mixture, because of its extraordinary power of haemoglobin, increases the volume of oxygen in the body by converting the sugar into glucose. When confronted by the enemy, this glucose is mobilized into the bloodstream from glycogen storage in the liver. It increases the heart rate, pupils of the eyes dilate faster and the hair stands erect. These reactions increase resistance to stress and enhance defensive abilities. Thus, warriors went berserk once equipped with biologicals and such aids. Alcohol and tobacco were a taboo to all warriors when preparing for war. It was realized that these have direct negative effect on performance. As narcotics, they suppress body functions and increase fatigue. They also reduce coordination leading to incoherence.

In the continuous search for efficiency and excellence in the military profession, people resorted to the use of extrasensory powers which were used in combination with natural preparations. It was believed that the latter gave the power to conjure up swarms of bees to scare off the enemy. The Kagoro people remained largely unconquerable throughout the precolonial period for this ability. Additionally, incantations essentially poetic in nature were recited and used in conjunction with herbal concoctions to make the opponent lose his memory so that his tactics would fail him and render him impotent.[8] It would seem that this activity was not without psychological effects on opponents.

As more destructive biologicals were experimented upon, man equally sought for and discovered the antidote for these. He discovered that some of the plant and animal toxins used to harm the opponent can be beneficial in protecting him in his desire for power over his fellows. Plants were discovered that could aid him in developing immunity against poison. Others helped in blood clotting and in the suppression of pain. Blood letting was also adopted as a method of countering the destructive effects of poison. This was done by making deep incisions on the affected part and the impure blood 'let out' with the aid of a horn that is opened at both ends. Thus, the increased destructive power of biologicals was matched by improvement in the treatment of victims. War injuries were treated by the use of anaesthetics, antibiotics and also shock treatment. Antidotes were prepared and eaten to develop immunity. These were often carried to battle in a sheep's horn and orally administered to the victim on the battle field. Antidotes were prepared from black paste made of carbon, shea butter and other materials like *Bridelia ferruguinea* and *Euphorbia hirta.* Preparation against arrow or spear poison was made by eating a small amount of the crushed heads of poisonous snakes such as vipers, crushed wasps and bees long before battle, so as to develop immunity.

Bridelia feruguinea (Kirni.)..irritates the respiratory leading to instant death.

The procedures for preparation were influenced by the sociocultural and religious beliefs of the community involved. In most societies of precolonial Nigeria, diviners, seers, spiritualists and traditional medical practitioners all featured in "military medicine." Arrow poison from plants was extracted by first placing the plant in cold water. It was then boiled and allowed to simmer after which the mixture was allowed to stand. This extract was filtered and used in smearing arrows and spears. During the period of preparation of arrows and spears, those involved had to abstain from sexual intercourse throughout the period to avoid spoiling the efficacy of the medicine. They also had to fast for at least two days. This is why in most precolonial Nigerian societies, during such periods, those involved had to withdraw to the bush to return only after all arrows were poisoned.

Warfare therefore forced peoples in Nigeria to research into medicinal plants and animals, their uses, and pharmacological content. The employment of these in warfare led to great advantages over other weapons of war as they were relatively mass killers thereby substantially reducing the military effectiveness of the enemy. Biologically based antidotes reduced mortality and became closely guarded military secrets so that their possessors would be perpetually on the offensive. This brought unity and co-operation between the different segments of society where military and military-related specialists contributed toward success in battle. Warfare therefore became closely related to the culture of society where development in military science took a cultural perspective. These culture-based developments in military science helped in promoting the cultural heritage of society and fully equipped individuals within it to live and meaningfully contribute towards its development. The modifications affected in biologicals from time to time for greater effectiveness in warfare indicates the cultural dynamism in precolonial Nigerian societies. This ingenuity, credulity and insatiable curiosity for excellence in military science helped in the development of peoples' potentials and, thereby, created opportunities for creativity and originality. It also helped in the preservation of societies' independence. The fact that the British invading forces were forced to study precolonial systems of warfare before attacking them indicates their effectiveness.

Military science in Nigeria stands to gain if research can be carried out into some of these biologically based strategies and techniques utilized in the precolonial era.

NOTES

1.   Achi, B., "Aspects of Military Technology in Nigeria before 1900." In Tarikh, Vol. 11, 1989.

2.   _____, "Arms and Armour in the Warfare of pre-colonial Hausaland." *African Study Monographs*, Vol. 8, No. 3, 1988, pp. 145-157.

3.   Lt. Col. A. H. W. Haywood, *Sport and Service in Africa*. London, 1962, pp. 63-64. See also A. N. Bello, (N. D.), *Gandoki*, Northern Nigeria Publishing Company, Zaria, pp. 1-10.

4.   Nelson, A., *Medical Botany*, E., and S. Livingston, Edinburgh, 1951, pp. 464-465. See also A. Sofoworola, *Medical Plants and Traditional Medicine in Africa*, John Wiley and Sons Ltd., New York, 1982, pp. 209-210.

5.   Wilson, J. R., *The Red Men of Nigeria*. Frank Cass and Co., Ltd., London, 1967, pp. 175-217.

6.   Afrigbo, A. E., "Towards a Study of Weaponry in Precolonial Igboland." Paper presented at the National War Museum Seminar on Nigerian Warfare Through the Ages, Owerri, 13-14 January, 1985, pp. 1-15.

7.   Heathcote, D., "A Hausa Charm Gown," *Man. Journal of the Royal African Institute*, 9, 4, 1974, pp. 620-624.

8.   Sofoworla, A., *op. cit.*, pp. 1-5. Also J. S. Boston, "The Supernatural Aspect of Disease and Therapeutics among the Igala," in the *Traditional Background to Medical Practice in Nigeria*, University of Ibadan Occasional Publication No. 25, 1971, pp. 7-12.

I wish to thank Dr. H. S. Isah, Department of Chemical Pathology, Ahmadu Bello, University Teaching Hospital, Zaria, Nigeria and Dr. I. Abdu-Aquye, Department of Pharmacology and Clinical Pharmacy A. B. U., Zaria, Nigeria for their input, analysis and professional advice.

Chapter III

Arithmetic in the Pre-Colonial Central Sudan

Ahmad Mohammad Kani

It is almost certain that when Prophet Muhammad implored people to "seek knowledge even in China" he was not referring only to what we now call the 'theological' sciences, because such sciences were not known at the time in such places. One is tempted to believe that the Prophet was referring to knowledge as a complex whole, that embraces all kinds of disciplines both 'divine' and 'secular.'

It was probably in response to various quranic injunctions that philosophers like al-Jabra, al-Hasan b. Haytham, Ibn Sina, Ibn Rushd, al-Farabi, al-Fakhr, al-Razi and al-Khawarizmi took it upon themselves to explore all avenues of knowledge and treat them not in a fractionalized form but as an organic whole to meet part of the overall objectives of Islam. Many philosophers of Islam were specializing in both the theological and 'secular' disciplines. It was not uncommon before the stagnation and subsequent decline of Islamic scientific investigations, in the sixteenth century, to find philosophers like Ibn al-Haytham, Ibn Sina and al-Razi combining mathematics with jurisprudence medicine and *Tawhid* (the science of the Unity of God) and physics with philosophy and sufism.

Ibn Khaldun recognized *ʿIlm al-Hisab* (Arithmetic) or numerology as part of the Islamic sciences, especially in relation to inheritance laws which required much arithmetical dexterity for inheritance to be justly divided among the rightful heirs such as the sons and parents and other relations like the nephews of the deceased (Ibn Khaldun, 1967).

The jurists required the use of fractions like 1/8, 1/4, 1/3, and 1/2 - hence the urgent need for the study of arithmetic. The involvement of "numbers, fractions, roots," in the division of inheritance had prompted Ibn Khaldun to consider "Inheritance Law" as a subdivision of arithmetic, (Hiskett, 1957).

It is worth noting here that qualities such as "truthfulness" "soundness" and "self discipline" have been closely identified by the Muslim philosophers with the study of arithmetic, thus demystifying the abstract notion of arithmetic and mathematical studies in general. The student of arithmetic according to Ibn Khaldun will eventually get accustomed to saying the truth and adhering to it in a logical fashion.

Within the Islamic tradition, another compelling factor for studying arithmetic is to grasp the system of collecting and distributing zakat (poor-dues) among the different grades of people who are entitled to it. Also the 'complexity' of the system of gathering the zaket in cash and kind compels the Islamic state to employ people well versed in arithmetic and accounting as zakat collectors. It was probably due to the above reasons that Ibn Khaldun innovated the term 'business arithmetic' as one of the branches of arithmetic. The object of such a branch, according to him was to cater to business transactions like the sale of merchandise. Another function of business arithmetic was to help in land surveying and similar enterprises that required the use of numbers. A cursory look at Ibn Khaldun's survey of the science of mathematics may reveal the fact that the development of socio-political institutions in Islam prior to the 16th century was parallel to the development of intellectual investigations in the fields of 'social' and 'applied' sciences.

With regards to the introduction of arithmetic to Nigeria, it is difficult to say with any precision when it was first introduced. Of course one is not suggesting that the concept of arithmetic had not existed before the introduction of Islam but this paper is primarily concerned with the type of arithmetic being taught now in both the 'secular' and Islamiyya schools in the country. One can assume that the introduction of arithmetic in Nigeria took place probably some time after the 11th century, the time when Islam is said to have been first introduced in the courts of the Mais (rulers) of Kanem-Borno. In Hausaland it is possible that the introduction of the science of arithmetic was simultaneous with the diffusion in the 15th century of theological, jurisprudential and Islamic law books such as *al-Mudawwana al-Kubra* of ᶜAbd al-Salam b. Sahnun (d. 858 A.D.) *Risalat Ibn Abi Zayd al-Qayrawani* (d. 995 A.D.) *al-Mukhtasar* of Khalil b. Ahmad (d. 1365-66 A.D.) and similar works on *Fiqh* (Islamic jurisprudence). The study of some branches of these sciences required basic understanding of arithmetic as pointed out earlier. Also the observation of certain religious functions and festivals like prayers and fasting and ᶜIdᶜ demand some knowledge of arithmetic by the ᶜUlama.' These factors and other related factors might have necessitated the inclusion of ᶜIlm al-Hisab in the curricula of education in Kanem-Borno and Hausaland.

It is evident that the study of astrology, geomancy, ᶜIlm al-Awfaq (the science of magic squares) ᶜIlm al-Huruf (the science of letter magic) and ᶜIlm al-Asrar (the science of secrets) demands some basic knowledge of arithmetical formulation. Documentary evidence at our disposal suggests that by the 17th century some ᶜUlama' (scholars) of Kanem-Borno were highly skillful in the sciences of ᶜIlm al-Awfaq, ᶜIlm al-Falak (astrology) and ᶜIlm al-Huruf. The overindulgence by some of the ᶜUlama' in such disciplines prompted the orthodox scholars like al-*Shaykh* Muhammad al-Wali to write a treatise in condemnation of the pursuance of such sciences for what he considered to be their diversionary and destructive aspects. According to the *Shaykh* the manipulation of these sciences was done at the expense of the most beneficial aspects of knowledge like the study of the Qur'an, jurisprudence and other subjects.

By the 18th century, the Borno kingdom became the most important centre of learning in the Central Sudan as far as occult sciences were concerned, attracting people from adjacent areas. So far the most renowned scholar in the field of occult sciences in the whole of the Central Sudan, Muhammad b. Muhammad al-Katsinawi al-Fulani (d. 1741) is reported to have studied the science of secrets.

However, it was at Mecca that Muhammad al-Katsinawi is reported to have begun writing one of his voluminous books on astrology entitled *al-Durr al-Manzum wa Khulasat al-Sirr al-Maktum fi ᶜIlm al-Talasim al-Nujum* (*the string of pearls and the quintessence of the concealed secret in the science of talisman and astrology*). And as his Egyptian biographer ᶜAbd al-Rahman al-Jaberti mentions, the book is a comprehensive work arranged with an introduction, five sections and a conclusion. According to his biographer, al-Katsinawi completed the work in Egypt in the year 1146 A.H./1722 A.D. Before his death in Egypt al-Katsinawi is reported to have written another book on the occult sciences entitled *Rahjat al-'Aafaq wa 'Idah al-lubs wa al-Ighalah fi ᶜIlm al-Huruf wa al-'Awfaq* (*the magnificence of the horizons and simplification of complicated and hidden (rules) in the sciences of secrets and magic squares*).

Before he finally moved to Borno and later to the Middle East, Muhammad al-Katsinawi is reported to have studied arithmetic and the sciences of letters and magic squares. He is also reported to have studied the science of secrets together with its "arithmetical and timing tools." The proper assimilation of these sciences by al-Katsinawi points to the antiquity of the occult sciences in Hausaland and beyond. Al-Katsinawi is also said to have excelled himself in Arabic grammar and logic confirming the notion which says there is a close association between the sciences of logic and grammar on one hand, and arithmetic on the other. Many students in Egypt are said to have been taught by al-Katsinawi. For instance the host of al-Katsinawi, *al-Jabarti al-Kabir*, is said to have studied under al-Katsinawi the sciences of *Awfaq* and fractions and numeration in addition to the sciences of letters and numerology.

It may be significant to note here that there is ample evidence to prove that the scholars of Hausaland and Borno were consulting coptic Solar Calendars in determining their economic-activities especially agricultural ones. The recovery of a book written probably in Egypt on agrarian activities, from Bauchi in 1973, points to the fact that some aspects of the agricultural sciences were being diffused in this area from time immemorial. The book which is copied in a Sudanic script contains many charts dealing with agronomic activities such as the right time of harvest, the various directions of the wind, time of germination and the seasons during which insects appear. A conversion table to lunar months is also made at the beginning of the book as a guide for the users of the chart.

It seems that some scholars in the Central *Bilad al-Sudan*, especially the area of Katsina, were well versed in the sciences of numerology and astrology. The recovery of some books from Katsina and other areas such as Borno by the late Professor M. A. al-Hajj and other researchers suggests that the scholars of Katsina were not only versed in these occult sciences but they were determined to defend the legality of pursuing such studies from the Islamic point of view (al-Hajj, 1968). The astrological and numerological works produced in these areas are also concerned with family problems, agricultural activity, trade and other forms of socioeconomic activity. In most cases the scholars resorted to the solar calendar to forecast and determine future events. However, the anonymity of the authorship of such books suggests that these sciences were viewed by some scholars with skepticism, as we will see later.

The 19th-century Jihad movement in Hausaland has been rightly described as an intellectual revolution which threw the door of academic pursuit open in all its ramifications (Kani, 1985). Education was a major preoccupation of the Sokoto Jihad. There is ample evidence to suggest that *Shaykh* ^cUthman b. Fudi was teaching both the simple and advanced arithmetic *(al-Yasir wa al-Gharib)* to his students. Another evidence of the incorporation of arithmetic and related sciences in the syllabus of the schools in 19th-century Hausaland is to be found with ^cAbd al-Quadir b. al-Mustafa who is reported to have studied medicine, astrology, arithmetic, logic and astronomy. He is also said to have studied *'Ilm al-Awfaq, ^cIlm al-Huruf, ^cIlm al-Rami* (geomancy) and *^cIlm al-Ajfar* (divination).

It is significant to state here that it was the 'official' policy of the Sokoto Caliphate not to indulge in the realm of the occult sciences. This policy was initiated by *Shayky ^cUth*man b. Fude and was emphasized by his son and successor Muhammad Bello. In his condemnation of the pursuance of such knowledge, Muhammad Bello is said to have gone to the extent of equating magical performance with astrology. However, despite Bello's condemnation, he himself had resorted to writing a book entitled *al-Kitab al-*

Kafi fi ^cIlm al-Jafr wa al-Khawafi (*the sufficient book in the sciences of divination and hidden knowledge*). The work involves complicated mathematical calculation aimed at forecasting future events.

Mindful of the 'official' policy regarding the 'prohibition' of disseminating the sciences of philosophy and the occult sciences such as astrology, ^cAbd al-Qadir b. al-Mustafa (d. 1864) tried to defend his position.

Arithmetical figures were manipulated by ^cAbd al-Qadir b. al-Mustafa with the intention of predicting the fall of the Sokoto Caliphate. He adopted the formula of *^cIlm al-Jafr* to arrive at what he considered the 'actual' date of the fall of the Sokoto Caliphate which he located in the sixteenth diagram of the world orders (*Abd al-Qadir*, NHRS). The process of arriving at the actual date of the demise of the Caliphate was done through addition, subtracting, division and multiplication of certain figures symbolized by specific letters. According to ^cAbd al-Qadir b. al-Mustafa, the fall of the Sokoto Caliphate would materialize as a consequence of one of the following: either by the appearance of the Mahdi ("the expected deliverer") or by the encroachment of "a vast power (British Imperialism?) which will bring down what has gone up and bring to prominence what has gone down." Coincidentally, forty years after ^cAbd al-Qadir's death, the British Imperialist powers invaded the Sokoto Caliphate.

To conclude, one is obliged to say that Nigerian historiography is lagging behind as far as the history of intellectual and scientific ideas is concerned. Most of the attempts made to reinterpret Nigerian history do not advance our knowledge beyond the political boundaries of historical investigation. Despite the availability of a great deal of literature on medicine, astrology, arithmetic and other related sciences, written in Arabic, Fulfulde, Hausa and other languages, little effort has been made to systematically study these sciences within the historical perspective. The intellectual output of the *^cUlama*, (scholars) in this area has been wrongly classified by our contemporary historians and social scientists under the rubric of 'mysticism.' A serious investigation into the literary output of the scholars of the Western and Central Sudan, however, may reveal the fact that these scholars had explored agricultural, medicinal, astronomical and mathematical sciences long before the advent of colonial rule.

REFERENCES

M. A. al-Hajj. "A seventeenth century chronicle on the origins and missionary activities of the Wongarawa," *Kano Studies*, Vol 1. 8. 1968.

Boyd, J. *The Caliph's Sister*. Chapman Hall, 1989.

M. Hiskett. "Material relating to the state of learning among the Fulani before their Jihad," *B. S. O. A. S.* XIX 1957.

A. Kani. *The Intellectual History of the Sokoto Caliphate* 1985.

Ibn Khaldn. *The Muqadimma*. Trans. Franz Rosenthal, New York, 1967.

Agsam al-Mirah wa Asbab al-Mirath wa Mawani^c *al-Mirath*. Northern History Research Scheme of Ahmadu Belic University, Zaria, collection hereafter to be referred to as NHRS. Ref. p. 70/1.

^cAdb al-Qadir b. al-Mustafa. *^cAbda' bi 'Ismi Kadiqi al-'Anwari*, NHRS, p. 25/5.

Kitab fi 'Ilm al-Falak. NHRS, p. 98/1.

Chapter IV

The Control of Water-Based Diseases in Colonial Northern Nigeria: The Case of Schistosomiasis

Gloria Thomas-Emeagwali

In this chapter, our major focus is on schistosomiasis or bilharzia. We are concerned with the dynamics of transmission of the disease and its major symptoms. We then proceed to identify areas primarily affected by the disease in Northern Nigeria during the colonial period. We are concerned primarily with the various control measures attempted by colonial authorities, measures which may be classified as ecological, chemical, biological as well as education-based strategies. We make recommendations for contemporary control measures in the light of the historical experience.

SCHISTOSOMIASIS

Schistosoma mansoni, Schistosoma japonicum, Schistosoma intercalatum and *Schistosoma haematobium* are the four main types of schistosome worms that infect man globally. It has been pointed out that the parasites themselves are of great diversity with much sub-species variation and that it is difficult to distinguish them with routine morphological techniques.[1]

It is the adult worms and the eggs they shed that are the principal cause of disease manifestation. Moreover the female converts the equivalent of her own body weight into eggs each day.[2] Egg counts reach as high as 6,800 eggs per gram of faeces in some cases.[3] The vectors of the schistosome worms are snails but the nature of interaction between the snail hosts and the parasites remains complex and not fully understood.

The dynamics of transmission and clinical features of the disease, bathing in streams contaminated with *Schistosoma mansoni*, led to acute shistosomiasis, after 50 to 60 days, in Brazil.[4] The widespread effects of washing, bathing and swimming in transmission sites of *Schistosoma mansoni* have been identified in the case of St. Lucia.[5] Moreover, the spread of irrigation projects and man-made water resources in zones of high incidence of schistosomiasis has led to a further spread of the disease,[6] particularly where such projects are not accompanied with the provision of adequate, safe and convenient domestic water supply for the generality of the population. Snail control by the use of molluscicides in water bodies remains one of the measures of schistosomiasis control.[7] But more significant is the provision of adequate, safe and convenient water. We shall say more on this issue in the course of discussion.

Some of the symptoms of schistosomiasis are abdominal pain, and bouts of bloody diarrhoea and fever.[8] But it has been pointed out that acute schistosomiasis has simulated cholera, typhoid fever, gastroenteritis and a wide range of diseases.[9] Studies on the appendix, brain and pancreas in 88 autopsies in Ibadan showed that the changes in the various organs were directly related to the intensity of schistosome infection.[10] Physical examination of human hosts revealed distended abdomen, clinical anaemia, lower haemoglobin and other such symptoms, in Uganda,[11] and spinal cord complications have been also identified in acute cases of schistosoma infection.[12] As pointed out by Weil and Kale the nature of the disease is related to the species of the schistosome worm that is the agent of infection: rectal bleeding and enlargement of the liver and the spleen for *Schistosoma mansoni;* abdominal pain and diarrhoea for *Schistosoma japonicum;* and urinary tract-related symptoms for *Schistosoma haematobium.*[13]

Patterns of Endemicity in Northern Nigeria

The Kagoro River in Plateau province was identified as a transmission site of schistosomiasis in 1929, following the infection of some Europeans through contact with that water body. *Schistosoma haematobium* were identified in the urinary samples of 98.03% of the school boys examined in Pategi in 1952, and Umaru Gana of the Medical Field Unit, Kaduna, found that no stream to the East of Pategi town except the Dzwafinni River was free of the aquatic snails associated with the disease.[14] Investigations carried out by colonial authorities in the primary schools in Ilorin, in the present Kwara State, found that Baboko Primary School had an infection rate of 68.8% and Oke Suna Primary a rate of 50.23% in that year.

In 1933 the Pathologist, Schistosomiasis Research, Zaria pointed out that schistosomiasis was highly endemic in Bida. The ova of *Schistosoma haematobia* were found in 32% of 5,323 males having a mean age of 25.52 years, whilst the ova of *Schistosoma masoni* were found in 17% of 4,349 males of a mean age of 26.92 years. The Pathologist also pointed out that of 3,880 persons on whom the *Fairley Intradermal Test* was performed, 73% gave a positive reaction.[15]

In 1933, *Schistosoma mansoni* infection was high in Sokoto, Jos and Kano and since urinary tract pathology seemed to be exceedingly high in Birnin Kebbi, Sokoto, Katsina, Kano and Zaria, we may conclude that in those areas *Schistosoma haematobium* was prevalent. In the valleys of the Niger and Benue rivers and on the high Jos Plateau, urinary schistosomiasis appeared to be less common, according to the Pathologist in charge of Schistosomiasis Research.[16] In the case of Bauchi, Maiduguri, Yola and Pankshin both *Schistosoma mansoni* and *Schistosoma haematobium* were common. As pointed out earlier, the vectors of the schistosome worms are snails of different species. Researchers identified *Bulinus tchadiensis* as the common vector of *Schistosoma haematobia* in Sokoto province, whilst *Planorbia pfeifferi* was identified in Sokoto and Bida, where there was the incidence of *Schistosoma mansoni*.

Schistosomiasis Control

It has been suggested that no single method of control can lead to the complete cessation of transmission of schistosomiasis.[17] In St. Lucia, the 92% fall in the level of water contamination and the subsequent control of the disease was achieved by routine mollusciciding as well as chemotherapy.[18] Aerial application of molluscicides as a single measure is reported to have had mixed results in the Sudan and Brazil.[19] Chemotherapy alone simply reduces the worm load in the hosts as well as the contamination of the environment but does not curtail transmission significantly enough.[20] It has recently been argued that vaccines against the disease may have to be supplemented by chemotherapy, vector control, environmental sanitation and other such measures unless the vaccines developed are considerably long lasting.[21] It would seem that a multi-pronged approach is the most effective method of schistosomiasis control. It is in the light of this that we now evaluate some of the methods utilized during the colonial period. We classify these into three broad categories, namely, ecological control methods, chemical-and biological-based strategies and public health education.

Ecological Based Measures

It was the declared intention of colonial authorities to establish waterworks and well-construction schemes in Kaduna, Kano, Zaria, Jos, Ilorin, Katsina, Maiduguri and Sokoto. This view was articulated as early as 1928.[22] For this purpose various investigations were planned, in terms of geological surveys, as well as preliminary inquiry into possible areas of well construction in the various localities. In the case of Ilorin Province it was observed that the greater part of the area was underlain by crystalline rocks and that this led to uncertainty in terms of the location of water.[23] Wells were to be dug at centers of population density in the case of Maiduguri rather than at intervals of five miles as earlier planned.[24] The program of well construction was not aimed at schistosomiasis control primarily but was rather envisaged as part of a general program aimed at relieving "the distress of the peasantry" with respect to the water supply.[25] As we shall see, however, this laudable objective was often unaccompanied by the funds necessary for actual implementation.

More directly related to the control of water-based diseases was the program of well management and maintenance. Several measures were suggested in that regard. Wells were to be periodically cleaned, and well collars and drainage aprons (sloping outwards from the collar for 8' or 10') were to be fitted on to wells. We should note, however, that this was to be done 'to all wells used by the European community and wherever else practicable.' It was recognized that a major source of water pollution was the use of dirty ropes, falling objects, the backwash of water from the field, and in some cases ulcerous legs, and it was advised that these issues should be attended to.[26]

Correspondence related to the funding of the waterworks and well-construction program is instructive however. It reveals that the sincerity of the colonial authority with respect to the project was more apparent than real. It also explains why the program was marred by premature policy shifts, uncompleted wells, and the reduction in the number of wells originally earmarked for some areas. Not long after the formulation of a fairly comprehensive policy for well construction in the Maiduguri region, the Secretary to the Northern Province directed that 'the matter should remain in abeyance for the present....'[27] In the Ilorin region, it was stated that only twenty-five of 279 wells would be actually constructed.[28] By 1952 it would be clearly stated that 'the water-supply scheme possesses no priority at the moment.'[29] Several unfinished wells were left behind by the ritual water supply teams, as pointed out by the Resident, Maiduguri in 1953. The fact is that the colonial authorities were adamant that expenses on the project were to be borne by the Native Authority and for this reason geological surveys and volumes of paperwork on the matter ended up as being purely academic exercises with little practical outcome. (The estimate for water supply in Ilorin, Emirate Division, Lafiagi-Pategi Division and Borgu Division was 45,000 in 1944.)

Biological and Chemically Based Measures

The use of molluscicides of plant origin was identified by colonial authorities as an effective measure against 'bilharzia carrying molluscs' and for this reason it was suggested that emulsions of Aduwa were to be applied to water bodies. It was believed that the latter was effective against molluscs in the water without affecting the potability of the water. It was recommended that the planting of the fruit in selected areas such as the vicinity of new wells, ponds and streams was an effective measure for schistosomiasis control. Aduwa trees were planted by the Agricultural Department in Katsina, Kwara and Sokoto.[30] It was hoped that this measure would also be useful against Guinea worm infection. We may note that the fruits *sapindus saponaria, Balantes aegyptiaca, Phytolacca dodencendra* (ended) as well as the pulp of *sisal* have been identified as potentially effective against the intermediate hosts, the active ingredients of the latter being *saponins*, which are lethal to cercariae.[31] *Aduwa* was believed to be the local name of *Balantines Aegyptica* the plant alluded to by Archibald in his influential paper.[32]

In an effort to reduce the breeding of the Anopheles mosquito, experiments were made with larva-eating fish.[33] We do not have evidence of similar attempts with respect to predator snails, a method which has been identified as an important control measure more recently. However, this particular control measure may not have been known during the period in question.

One of the important chemically based strategies, involved the use of *copper sulphate* as well as *Santorin* in the rice fields of Pategi, which were identified as infected with cercariae. We may note here that there was a keen awareness of the spread of the molluscan hosts with irrigation projects and a combination of 'mass treatment and anti-cercarial measures' was recommended. Additionally there were occasional laboratory tests of urinary and faecal samples of school children in selected areas, a measure which was utilized during the last decade of colonial rule in particular. Laboratory tests provided fairly specific data on the clinical manifestation of the disease and the overall rate of infection as illustrated earlier in this paper. In the colonial period it would seem that chemical and biologically based strategies were more effectively carried out than the well-construction and water works program, comparatively speaking, although in some cases applications were sporadic and also starved of funds. At any rate the former measures never had the capability to eliminate a major cause of water-based diseases such as - *schistosomiasis*, namely, inaccessibility to a clean and regular water supply. In the main, chemical and biologically based strategies were secondary and the provision of safe water per person per day, on a regular basis, could only be achieved in the context of a successful water works program and well-digging. As we have seen, however, the latter remained elusive throughout the colonial period in most communities.

Health Education

It has been pointed out that few rural inhabitants understand the relationship between the quality of water, waste disposal methods and water-based diseases such as *schistosomiasis*.[34] We may also add that the number of rural dwellers who comprehend the relationship between the parasites and their vectors, as it relates to the transmission of the disease, is even less. It is in recognition of this factor that the colonial authorities embarked on a program of health education in the period of focus. One of the strategies associated with this exercise was the preparation of pamphlets entitled 'Nigeria's Health Service has this message for you in the Northern Provinces.' Under Section II of the pamphlet on *Water Supply*, reference was made to *schistosomiasis*:

> In the case of Schistosomiasis which so many people suffer from, the infection is contracted from bathing in stagnant pools e.g. borrowpits...People must not urinate in any bathing place.[35]

Pamphlets written in Hausa, in poetic form, were also composed.[36] See Appendix. Judging from the reaction of the translators approached by the present writer, the Hausa used was Sokoto Hausa and in its present form was generally difficult to comprehend. This may be due to the various changes in syntax and vocabulary that the language has undergone since 1938, however, and we await more specialized opinions on the issue. It would seem, however, that the verses contained several misinformed perceptions about the disease itself.

Teachers and prefects were requested to inform students of the cause of the disease and to warn them against exposure to contaminated water.[37] (Routine laboratory testing of urinary and faecal samples of school children may be seen as auxiliary to this aspect of the program.) Archival sources suggest the absence of coordinated supervision with respect to implementation of the program.

The employment of Sanitary inspectors in the various communities may be seen as another aspect of the Health Education policy. The latter were requested to send copies of their inspection notes to residents. Moreover, they were to visit private and public premises and make specific suggestions for improved sanitation in specific areas. Their activities were not confined to the inspection of water bodies but rather to all public and private premises, a scope which was rather wide-ranging in the light of the relative shortage of staff available for the exercise.[38] What seems clear is that financial constraints inhibited adequate staffing and compensation of the Inspectorate which often complained of overwork.

Implications for Contemporary Policy

British colonial authorities were associated with attempts at the control of water-based diseases such as schictosomiasis in colonial Northern Nigeria. Some of the programs were laudable enough on paper, but, given the absence of financial commitment to such programs, the gains made were at best minimal, and hardly commensurate with the vast amount of file-passing and paperwork associated with the exercise. Colonial authorities found little difficulty in unfolding grandiose plans for well-digging and waterworks in the various communities, but these plans were in most cases not matched by effective and practical results. The fact remains that, by virtue of its underlying structure and orientations, a colonial state is hardly ever welfarist and humanitarian in its objectives and strategies. The British colonial state was an exploitative coalition of forces whose prime objective was the relentless accumulation of the surplus labour of its subjects. It was expected, therefore, that colonial health policy would be superseded by more effective policies and an improved record of health care vis-à-vis water-based disease, in the context of the post-colonial and post-independence structure which emerged after 1960. This expectation, however, is yet to be fulfilled, judging from the contemporary statistics as they relate to *schistosomiasis* and other water-based diseases.[39]

Contemporary state policies are unduly preoccupied with prestige projects and various unrelated strategies for the consolidation of minority and sectional interests. Additionally the contemporary debt crisis which implies astronomically high rates of debt payment and the removal of subsidy on health and other sectors, do not augur well for the eradication of these diseases. In the light of the historical experience there is the need for a rational budgetary allocation in the interest of disease eradication, particularly since there is an increase in transmission sites; a well coordinated program for chemical and biologically oriented control in the light of more recent scientific discoveries; greater and more effective utilization of mass education techniques and devices in the context of a holistic health education policy; and, most importantly, an overall policy which is less preoccupied with verbal declarations than actual accomplishments and achievements on the ground. History will not absolve the present post-colonial generation if it fails to make concrete and identifiable contributions towards the eventual eradication of diseases such as schistosomiasis.

50

NOTES

1.   Newton B. and F. Michel (eds.), *New approaches to the identification of parasites and the parasites and their vector.* Schwabe and Co. 1984.

2.   Coles, G. C.. "Recent advances in Schistosome biochemistry," *Parasitology*, 89, 1984.

3.   Ongom and Bradley, "The Epidemiology and consequences of Schistosomiasis mansoni infection of the West Nile, Uganda" in Transactions of the Royal Society of Tropical Medicine and Hygiene, Vol. 66. 6, 1972.

4.   Naves J. *et al.*, "Spinal chord complications of acute Schistosomiasis mansoni" In *Transactions of the Royal Society of Tropical Medicine and Hygiene,* Schwabe and Co., Basel, 1984.

5.   Jordan, P. *et al.*, "Control of Schistosomiasis mansoni transmission by prevision of domestic water supplies: A preliminary report of a study in St. Lucia." *Bulletin of the World Health Organization,* Vol. 52.I, 1975, pp. 9-20.

6.   Sellin B. *et al.*, "Essai de lutte par chemotherapie contre schistosoma haematobium en zone iriguee sahalienne au niger," in *Medecine tropicale,* Vol. 46.I, 1986.

7.   Lyons, G., "Schistosomiasis in N. W. Ghana," *Bulletin of the World Health Organization*, Geneva. Vol. 51.6, 1974.

8.   Ongom and Bradley, *op. cit.*

9.   Neves *et al., op. cit.*

10.   Edington *et al., Society of Tropical Medicine and Hygiene* Vol. 69.I, 1975.

11.   Ongon and Bradley, *op. cit.*

12.   Naves *et al., op. cit.*

13.   Weil and Kale, "Current Research in Geographical aspects of Schistosomasis," in *Geographical Review*, Vol. 75.2.

14.   Gana, in NAK ILORPROF 174, "Guinea worm in Nigeria," *Schistosomasis, Prevention of* 1952.

15.   Walter in MINPROF 498, *Schistosomiasis in Bida.* 1933.

16.   *Ibid.*

17. Jordan, *op. cit.*

18. Barnish *et al.*, "Routine focal mollusciciding after chemotherapy to control Schiestosome Mansoni in Cul de Sac Valley, Saint Lucia," in *Transactions of the Royal Society of Tropical Medicine and Hygiene*, Vol. 76.5, 1982.

19. Amin *et al.*, "Assessment of Sudanic Snail control program," *Annals of Tropical Medicine and Parasitology*, Vol. 76.4, 1982.
Barbosa F. and D. P. Costa, "A long term Schistosomiasis control project with Molluscicide in a rural area of Brazil," Vol. 75.I, 1981, pp. 41-52.

20. Jordan, *op. cit.*

21. Butterworth, A., "Immunity in human schiastosomiasis," *Acta Topica*, 44. Suppl. 12, 1987, pp. 31-40.

22. Firmantle, in MAIPROF 555, *Waterworks schemes in Northern Provinces, Wells control of*, 1928.

23. Jones, in ILORPROF 1502, *Water supply ILORIN Division*, 1932.

24. NAK MAIPROF 2716, 1937.

25. NAK ILORPROF 1502, 1941.

26. NAK MAIPROF 555, 1937.

27. NAK MAIPROF 964, 1929.

28. NAK ILORPROF 4007/S.6, 1944.

29. Scott, R., in MAIPROF 1968, *Water supply Potiskum Division*, 1952.

30. NAK ILORPROF 174, 1934.

31. Mc Cullough F. *et al.*, "Molluscicides in Schistosomiasis control," in *Bulletin of the World Health Organization*, 1980, pp. 681-690.

32. See NAK ILORPROF 174, 1934.

33. NAK Min. of Health MED 1689.

34. Bayard *et al.*, "Rural water supply and related services in developing countries - Comparative analysis of several approaches," *Journal of Hydrology*, Vol. 51, 1-4, 1981.

35. NAK MINPROF 1007, 1947.

36. See Appendix.

52

37.   NAK MINPROF 497, 1933.

38.   Thomas-Emeagwali, G., "Dracontiasis in Ilorin, Nigeria 1915-1960" mimeo 1989a, Ilorin.

39.   Edungbola, L., "Water utilization and its health implications in Ilorin, Kwara State, Nigeria," in *Acta Tropica*, Vol. 37.I, 1989; "The distribution of Dracunculiasis in Nigeria: a preliminary study," in *International Journal of Epidemiology*, Vol. 17.1, 1988. pp. 101-106.

APPENDIX FOR CHAPTER FOUR

Wakar Chiwon Bilharziya

1. Bayan wagga waka akwai wata ga waka ta bilharziya.

2. Duk mai yin fitsari gami da jini shi adda bilharziya.

3. Alkwate addalilin ta halbi mutum shi zamo da bilharziya.

4. Alkwato nada halin munaffukai gudu nesa in kaga ta.

5. Wadansu irin mutane suna in ta baibai da ba hankula.

6. In ga kasa mutum yanzu yattake ta kafa ta rure masa.

7. Alkwato na da chiwo guda babba gun zuchchiya kunjiya.

8. In ta sami inwar mutum sosai halbinta ya tabbata.

9. Sai kaga ga kuraje tuli ga jiki na mutum shina chichchira.

10. Sai tayi kwai ta haife chikin ruwa ga yayan ta na zabura.

11. Kowa ya sha da su babu tafsawa gwaiwa ka samu nasa.

12. Diyanta chikin ruwa sai su halbi mutum shi zamo da bilhazita.

13. Kullum sai fitsari shikai da jini zarafi shi kare masa.

14. Mai safarar shiga ruwa tabki ko gulbi shi dau anniya.

15. Diyan ta suna chiki sai su halbi mutum beli shi kulle masa.

16. Diyan ta kasa ana gyambuna manya su dade ba sun warkuba.

17.   Diyan ta kasa mutum na hukashi zamo aiki garai ya rasa.

18.   Diyan ta chikin ruwa masu halbi suka kunji zamba tata.

19.   Kowa ya samu chiwon ga sai shi tafo gun Dokta mai anniya.

20.   In ya bashi yasha shi warke sai shi tafo gida lafiya.

21.   Chiwon nan da yazzo ijib alkaryar su ta firgita.

22.   Sunan mai garin nan Fu'adu shi yabbidi magani yarrasa.

23.   Sun rasa magani sai ga Dokta da yazzo sun ka warke duka.

Source: Ilorprof 4641, 1938, National Archives, Kaduna, Nigeria.

Chapter V

Metallurgy in Northern Nigeria:
Zamfara Metal Industry in the 19th Century

Nurudeen Abukakar

This chapter discusses findings on Zamfara metal industry in the 19th century. We focus on the processes of ore identification, mining, smelting, and smithing and the distribution of the finished products. We also focus on iron and other metals employed in Zamfara, which was by the beginning of the 18th century a power to be reckoned with in Northwestern Hausaland.[1] Even though it was hemmed in by rival and powerful states on all sides except the Southwest, and as such missed the influence which others gained by being absorbed into the wider West African World, it had managed to become one of the most important states of Hausaland.[2] For, according to Smaldone:

> ...after the brief interruption of the Jokun into Hausaland in the mid seventeenth century, and the resurgence of Borno in the early eighteenth century, it was Zamfara that emerged as the paramount local power among the Hausa states. By the mid eighteenth century, after a series of wars with Kano, Katsina, Gobir and Kebbi, Zamfara had become the dominant state in the Rima River Valley.[3]

However, both the process of its emergence and its extent are not very clear and all we have are generalized descriptions of its size and extent. We are told, for example, that at the height of its power, it extended from Sabon Birni in the north to Kwiambana in the south; from the rocks of Muniya, Rubu and Duru and the stream Baban Baki in the east, to the River Gindi in the west.[4]

Garba Nadama, who has examined Zamfara's history closely, opined that by mid-18th century, while the hub of the territory "lay within the areas covered by the major tributaries of the Rima river,"[5] the state has expanded up to the Rima area in the north. Despite the lack of "a clearly demarcated boundary."[6] Zamfara's authority is acknowledged by the present-day villages and towns west of Dajin Rugu, its edge being the boundary between Kasar Zamfara and Kasar Katsina. The southern limit of Zamfara is said to have "run along the Valley of Gulbin Zamfara, westwards to Girkau town near Kebbi in Rikon Kebbi."[7] To the northwest its limit was marked by Dajin Tureta or Dajin Gundami "running all the way from the north-western corner of Zamfara along the areas of Bakura, cutting across the middle of the Sokoto river down to the river Ka and merging with the dense forest to the south"[8] Caliph Bello uses the term Zamfara to refer to the conglomeration of the land and people who were inhabiting the eastern and southeastern part of metropolitan Sokoto.[9] By the beginning of the 19th century, the whole area described above was encompassed in the Sokoto Caliphate and formed part of its metropolitan section.

The Political Economy of Zamfara

Zamfara, like most Hausa states, was composed basically, of two classes of people, namely the *Sarakuna* and the *Talakawa* the ruling elite and peasants, craftsmen and merchants of different rankings. Before the Jihad there was no significant group of Ulama or scholars in Zamfara. The political ruling class derived their authority from forceful integration of the various communities. By right of conquest they extracted surplus from the various productive groups in Zamfara.

The social formation in the Caliphate (of which Zamfara was a part) comprised an assortment of modes of production. These are the feudal, whose tendency was most dominant and there was also the slave mode of production. In lesser significance as a mode but integral to both we find petty commodity production and exchange, both local and long distance.

The method of forcefully extracting surplus was primarily through tributes in all.[10] Thus, not only did the *talakawa* pay taxes they also formed a pool of labour from which the rulers got free labour through the institution of communal labour or *gayya*. In this system, the rulers employed the offices of influential farmers and chief members of guilds to mobilize the peasantry (*talakawa*) and collect taxes. Slaves were owned largely by the ruling elite and to a lesser extent by wealthy and influential merchants and craftsmen. The slaves were obtained largely through wars, even though slave trading was a conspicuous aspect of the system.

Thus, from the tributes forcefully extracted from peasants, craftsmen and merchants, the ruling class was able to maintain and consolidate its rule. It was able to maintain a large retinue of officialdom, wage wars, and capture slaves who were in turn invested in production especially in the royal farms and to a lesser extent in commodity production.

The organization of production was based primarily in the *Gandu* system of cooperative production and to a lesser extent on the household system both in agriculture and industry. Under the *Gandu* system, members of an extended family collaborated in production. The proceeds formed a common pool from which members drew for activities such as weddings and naming ceremonies and which they fed from during certain months of the year. While separate households may undertake independent activities at agreed days of the week or hours of the day, *Gandu* work was the primary one. The household system consisted of a man, his wife, children and other dependents. Each household was taxed and it was the *Gandu* that must send some representatives to the *gayya* whether it was to the numerous royal farms, or to rebuild or extend the city walls or to clear forests for new farmlands. Agriculture was the backbone of the society. The *Bukin dubu*, a ceremony organized at the attainment of a thousand measures both in agriculture and spinning, was geared toward maximum production. Thus, not only could Zamfara feed itself but grains were exported to the drier north. Even though Zamfara never attained the fame of Kano in the exportation of cloth, its cotton farms had been noted right from the 16th century.[11]

Zamfara definitely excelled in cotton weaving.[12] Agriculture in Zamfara also formed the source from which most of the crafts drew their raw materials. Agriculture, the crafts and commerce thus formed a fully integral economic system in that region.

The Iron Industry In Nineteenth-Century Zamfara
Training A Smith

Generally, hereditary occupation was the rule in Hausaland.[13] The ideal situation therefore, was for a child to "inherit" the trade of his father. There are a lot of Hausa sayings which stress this fact. Two examples are "*kyaunda ya gaji ubanai*" (the preferred thing is for a son to inherit his father's trade), or "*In dai ya tsari gidansu, ba dan banza ya tafi yawo ba*"[14] (If he protects the tradition of his family, he is not a useless boy who wanders away). The practice in the blacksmithing industry, like most others, was for a son to "inherit" the trade of his father. In the same spirit, a child usually learnt from his father or a close relation. The general method of training was by participation and observation under very close supervision.

The age at which children started learning was not fixed. In the series of interviews conducted, the blacksmiths cited the ages as being between six and ten years.[15] The essential factor, however, was for a boy to be strong enough to be able to wield the hammer. Initially, the child started experimenting with small objects in the absence of the master. These were knives, sickles and a kit consisting of a point, a scraper and a tweezer, used in removing thorns from under the feet.

The training of the blacksmith was very arduous. The difficulty in working wrought iron was often not properly comprehended and described by a number of scholars because these obtained their impressions from watching the working of the imported cast iron. The bloom or wrought iron consists of iron particles loosely adhering to one another, making them more compact and therefore less brittle as working implements. The bloom spreads while being hammered. It needs considerable expertise to achieve this.

On rare occasions, the blacksmiths took in outsiders as apprentices. This also obtained in other professions in Hausaland as has been mentioned earlier.[16] The problem an outsider (i.e., one who did not inherit the trade) faced was whether he had the patience to build up enough stamina. Outsiders were first stretched to the limits of their physical endurance.[17] Outsiders, however, took on the trade at a more ripe age. An outsider could move from one master to another in different regions specializing in the use of different implements.[18] An enterprising outsider could therefore be more exposed in the trade. This was however the exception, just like taking in an outsider was an exception in the first place.

Mining The Ore

The widespread occurrence of surface and subsurface iron stone has been described as "a notable feature of Sokoto province."[19] These stones occur principally in lateritic crust or capping on the rocks, in some places forming "thick sheets extending continuously for many square miles."[20] These iron stones "vary greatly in appearance, hardness and composition."[21] They in some places constituted iron ore, "the principal form being limonite."[22] Haematite is also available but more rarely, "chiefly in surface exposure where the iron ore has been dehydrated through isolation."[23] Another iron oxide found was gothite.[24]

While the differences in the mineral composition of the above-mentioned iron oxides are real and may be of academic interest, in practice this is hardly so. In reality not only could traces of one be found in the other e.g., traces of haematite in gothite,[25] but they often merged together. These are called banded iron formations.

Informants mentioned nineteen mines where ores were obtained in Zamfara land during the 19th century. By the late 19th and early 20th century Dutsin Disa in Tureta where smelters converged from all over Zamfara was the most popular. The type of ores used in Tureta were found to contain 41.25% iron.

The blacksmiths had different names for the ore types in different parts of Zamfara. The names were based either on the physical characteristics of the stone, e.g. *Anta*, which means liver-like, or *Kwan Kiffi*, fish egg, or based on the return got after smelting. In this instance good ores, that is those that yielded good returns, were called *Hanore* and bad ones, *Kiluji*.[26]

Ore identification was based especially on physical characteristics of color, luster, hardness, weight, and so on. There is a consensus that stones that would yield iron can be chewed. Most blacksmiths interviewed claimed that they "inherited" the mining sites, and that they mined where their predecessors had mined. Examples in this respect are Dutsin Disa in Tureta district and Dutsin Gwabro in Maru. Tureta had by the 19th century become the meeting place for smelters from different parts of Zamfara. It was pointed out by blacksmiths that ore identification was based on experience. Thus when a less experienced blacksmith found earth which he considered to be ore bearing, he collected some and referred them for further scrutiny by more experienced blacksmiths.[27]

Prior to the jihad, due most likely to the importance of iron to the society, blacksmiths never needed to seek permission from the Sarkin gari,[28] the political head of the town. They only bade him farewell, as a sign of courtesy, when departing from the town for the *zango*. With the onset of smelters crossing political boundaries, however, for security and economic reasons the movement of the smiths began to be monitored.

The equipment needed to mine ore consists of a variety of strong hoes and axes (*Magiri, Gatari, Dandurusu*), ropes made of the bark of the *kuka* or baobab tree (*Adamdonia digitata*), (Baobab) tree saddle bags, containers used in hauling the ore from the pit and donkeys, their only means of conveying the ore from the mine to the *zango* or camping site.

At the ore-bearing site, miners dug pits which looked like wells or dye pits. While the depth reached may not be known and varied they generally made efforts to reach the seam. Miners were usually dressed in leather loin cloth, if at all. The workforce normally consisted of the very strong and able-bodied members of different family groups. The work force in each group therefore consisted of the leader (the oldest active member of the family), his servants and journeymen, slaves if any, and other able-bodied members of his family.

Though the actual mining was done in independent groups, the whole operation was collectively done. They all departed the *zango* for the mine at a signal from the *sarkin zango*. In Maradun the signal involved the sounding of an iron gong. At the mine, work was carried on throughout the day. At nightfall, the ore was packed and loaded on donkeys. The departure time from the mine was announced by the iron gong. This routine they continued until they had mined enough ore. The distance between the *zango* and the mine was often considerable. Quite a number of the donkeys died before the mining operation was finished.

The Zango in Zamfara

Zango in Hausa means a camping place and it is usually associated with long-distance traders. Trade links go as far as the 13th century.[29] By 1450 Hausa traders had started venturing into the West African sub-region.[30] Trade between Hausaland and the Jukun state of Wukari was definitely on by c. 1500.[31] The *zango tama* was not a fixed or permanent place. Not only could there be several *zango* in a state, the smelters were not bound to any particular one. They could move from one *zango* to another depending on convenience. Nineteen of these sites were mentioned by informants.[32]

The smelters went to the *zango* with all their smithing equipment and were accompanied by some of their wives who did the cooking. By the 19th century, one may also find traders in the *zango*. Three categories of traders accompanied the smelters to the *zango*. There were those blacksmiths who may not know how to smelt or who might not be allowed to do so. These purchased the contents of several *murhu* unworked. They then worked on them and made *kunkuru* enough for different *ma'aikaci*. Another category of blacksmiths purchased the *kunkuru* and fashioned them into ready-for-use implements without the wooden handles.[33] Then there were traders in the proper sense of the word called *makeran kotato*[34] who purchased the implements for sale in markets distant and near.

Furnace and Tuyere Construction

We learned from informants that after the *zango* has been physically erected, the miners proceeded to the second undertaking, which is mining the ore. When they had stock-piled enough ore, the smelters then proceeded to the construction of the furnaces.

The furnace type employed in 19th-century Zamfara was the medium-height shaft. There was understandably no fixed measurement of the furnace. All the measurements were given in estimates and averages. Thus, the smelters estimated the normal height of a furnace to be up to the chest of a fully grown man.[35] It was also often estimated at three times the circumference of a human head.[36] Its width was such that it could almost be enfolded in a man's arms. The depth of the *murhu* from ground level to the deepest point was estimated as knee deep. The furnaces generally taper slightly from the base to the top. They were equipped with spy holes at about twenty centimeters from the top. The style of construction and the period of construction does differ. Thus, while it was built in just a day in Talatar mafara, in Sokoto, it took three days. Whereas it was constructed piecemeal in some places, in Talatar mafara it was worked at continuously until completed. In Gummi where it was also constructed continuously till completion, it was constructed in a spiral design. The furnace was considered the property of the *Sarkin zango* in Anka and Maradun. It was the *Sarkin zango* that supervised its construction. Individuals, however, made their *tuyeres* (*dodoba*).

Individuals made between thirty and fifty *muhu* (literally, a hearth, but here the reference is to content of one) using only a single furnace. In some cases, loops made of iron were made and used as an additional protection to prevent the furnace from bursting, something that often occurred. The furnaces were moved with ropes made from the bark of the *kuka* or baobab tree. The ropes were passed underneath with the aid of a stick. They were then carefully lifted by strong men and taken to the new location.

Kurt Krieger[37] gave the actual measurement of a furnace which he contracted to be constructed for him in Tureta. Its height was one hundred and forty centimeters (140 cm.). Its diameter 100 cm. at the base while it measured only 80 cm. at the top, showing that it tapered. Its thickness, though irregular, was estimated at 4 cm. Spy holes were made slanting from the outside to the inside, at the regular distance of between 25-35 cm. either individually or in groups of three situated 20 cm. below the upper rim. These were often made facing the points of the compass.[38] The *tuyeres* employed were semi-cylindrical, eight in number and measured 55 cm. long. The inner cavity had a diameter of 9 cm. and the walls 4 cm. thick. The diameter of the *Murhu* was between 60-70 cm., and its depth was 25 cm. Another hole was dug in the first one. Its diameter was 30 cm. and had a depth of 25 cm.

In constructing the furnace and *tuyere*, effort was generally make to get fine and slimy soil. Termite hills, referred to as *kasar kunkuwa*, were however preferred. In Maradun it was mixed often in a specially constructed pit, for between three and seven days. In Gummi it was left for fourteen days, while others just left it alone putting samples of all the grains they produced inside until it started sprouting. Others mixed the soil everyday. In the soil they added dried grass, hair removed from the skin of animals by tanners, and ground pods. All these were in order to make the furnace structure strong such that it not only withstood the considerable heat generated during smelting but in such a way that it also could be employed several times.

An integral part of the furnace was the *tuyere*. The *tuyeres* were constructed after the furnace had been finished. While only one furnace might be adequate for a series of smelting operations, for each a new set of *tuyeres* had to be provided. The number of *tuyeres* differed from seven to eight. Two methods of *tuyere* construction obtained in Zamfara. In the more commonly practiced, a stick was placed on a flat ground. Mixed earth that had been specially prepared was then moulded into it. The stick was carefully removed before the earth dried. In this method practiced in Gummi and Tureta among others, a semi cylindrical *tuyere* was obtained. In Talatar mafara the earth was moulded on the ground and pierced with a stick.

The *tuyeres* in Zamfara served a dual purpose. They were in the first place ventilation holes for the natural draught since no bellows were employed to induce draught. Secondly, even though they made provision for collecting slag at the bottom, slag was still tapped via the *tuyeres*. This explains why they had to employ new sets of *tuyeres* for each smelt. The *tuyeres* got blocked by the slag.

The duration of the smelt differed from place to place depending on several factors, such as the height of the furnace, the extent of tapering, the number of charges per smelt and the amount of ore vis-à-vis the fuel. The duration also depended on the strength of the wind since no bellows were employed. It also depended on whether it was wood that was employed as fuel or a mixture of wood and charcoal. Generally, however, the smelting process started in the evening, at about sunset. It lasted until the early hours of the morning.

In smelting the furnace was first assembled. The *tuyeres* were arranged in a circle at regular intervals, cavities facing the ground a one fourth of their length extending beyond the edge of the hole.[39] The longer section of the *tuyere* rested on the ground and was covered with earth, care being taken to make the earth cover at the circumference of the lower part of the furnace.

Ore was then poured in the furnace. The measure used, not only differed in size but also in the numbers employed per charge, depending on the size of the furnace. Thus, in Sokoto it was four measures of ore followed by wood. This was repeated four times.

Meanwhile, the ore that had been brought from the mine was worked upon before it was employed in smelting. The ore was puddled into fine particles. This was achieved by breaking the bigger pieces using mallets and stones. It was the puddled ore that was measured and used in smelting.

To make a bellow, the blacksmiths found a goat's skin which had no holes in it. The skin was treated with acacia liquid and left to dry. A piece of wood carved in the shape of a hoe was make, one end being very large and the other narrow. The large end of the hoe-like wood was connected via a metal pipe or nozzle. Two strips of wood which formed the handle were made and sewn on to the wide end of the skin leather. To the wood was attached another piece of leather through which the bellower placed his hand. This made one bellow. The bellower employed two, one in each hand. A "valve" or "door" was made of red leather and put where the bellower held the bellows, so that air could enter. When the bellows were opened, the valve opened, and when they were closed, the valve closed. To operate the bellows, the bellower employed both arms simultaneously in an opposing movement. Thus, while one arm was being raised and the hand opened, the other was being lowered and the hand closed. When the hand opened in the upward movement, air was sucked in. This air which got trapped by the closing of the hand in the downward motion, was forced through the nozzle into the *bakin wuta* and the hearth. By constantly repeating the movements simultaneously, the hearth was supplied fresh air for combustion.

The Guild of the Blacksmiths in Zamfara

The guild of the blacksmiths was a formal cooperative organization the authority of which was accepted by every blacksmith, sanctioned by the state and respected by the society at large. Its essential function was the promotion of the corporate interests of the blacksmiths. The guild of the blacksmiths, probably the best organized among the crafts' organization, was headed by the *Mazuga*, the title of the leader of the blacksmiths in Zamfara.

The guild of blacksmiths in Zamfara was structured as follows:

Households	(Made up of people with the same economic source as pool)
Workshops	(Made up of household members usually related by blood)
Lineages	(Made up of workshops forming a cloister in a ward)
Wards	(Many lineages in a single section of a town)
Mazuga	

In metropolitan Zamfara all the blacksmiths were directly under the control of the *Mazuga* of the towns. The *Mazuga* was assisted by wardheads of the wards (*shiyoyi*) occupied by the blacksmiths. The leader of the blacksmiths in each ward was appointed by the *Mazuga* from a particular family or a select number of families. Like the office of the *Mazuga*, the office of the wardhead was also hereditary. Under the wardheads were the leaders of lineages. A lineage would thus consist of members related by blood, divided into different households, but tracing their history to a common ancestor and constituting a substantial membership of a ward. A lineage would be headed by the oldest active member. Under the lineages were workshops that often constituted more than a household. A household was the smallest independent unit, there were several households in a single compound.

The household was made up of a man and his family that opted out of the gandu system. The *gandu* system was a sort of economic cooperative unit comprising members of an extended family where they pursued the same economic activity jointly. For some parts of the year they 'ate' from a common source; some cultural activities were also financed from the *gandu*, e.g., marriages and naming ceremonies. Meanwhile members would still have some independent source of income in the same profession which they could pursue at some agreed time of the day or days of the week when the *gandu* work was not in progress.

The blacksmiths did not formally pay tax during the 19th century. Because of their technical competence, they were seen as indispensable to the state and the society at large. They were not only exempted from paying tax, they were also exempted from all kinds of forced labour which other commoners were subjected to. The organization of the blacksmiths, though subsumed like others, was given more latitude in the society than any other professional group.

For the state, the blacksmiths annually, by the advent of the rains, collected, through the *Mazuga*, a number of hoes for onward transmission to the *Sarki* or Chief. The collection followed the structure drawn up above. Meanwhile, at each level in the hierarchy the person responsible retained a portion as the rewards of his labour. This obtained up to the *Mazuga* who benefitted substantially from the collection exercise.

The same method was employed almost annually also for collecting armaments during the dry seasons, especially if there was an imminent campaign in sight. The blacksmiths manufactured the iron gates and other iron implements needed by the state. The blacksmiths produced the heavy hoes and axes used in clearing bushes and forests (*Birgi*) for new royal farms (*Gandun Sarauta*). It was in recognition and appreciation of these efforts that the blacksmiths were exempted from the payment of taxes and any other kind of forced labour as earlier mentioned. The blacksmiths had optimal practicing conditions in terms of accessibility of raw materials and minimal interference from political office holders. There was a kind of mutual respect between the blacksmiths and the political authority. The blacksmiths did not consider their various contributions to the state as exactions. It was the minimum they could do to a society that held them in such good regard.

Though all sectors of the society depended on the blacksmiths, the relationship between them and the society was often portrayed as if the society consisted only of farmers and blacksmiths. *Dan inganta noma shi yasa kira take da muhimmanci* (it is because smithing strengthens and reinforces farming, that is why it is an important industry). The relationship between the blacksmiths and farmers was certainly the most intense. No other group depended on the blacksmiths as the farmer did. Butchers, tanners, cobblers, barbers, housewives and even warriors all depended on the blacksmiths for the effective pursuit of their trade.

From the advent of the rains until harvest, the relationship between the blacksmiths and the farmers was hectic. It was more personal than legal, and therefore needed extra legal methods to maintain the equilibrium needed for a harmonious relationship especially as blacksmiths often worked for farmers throughout the rainy season without receiving payment, waiting till harvest. Even at harvest there was no legal prescription of what amount to pay. Payment was based on affluence, the harvest, whether good or bad, and the other commitments of the farmers. On the other hand, the farmers also contributed grains when the blacksmiths went to the camp. The arrangements were complex. Often families attached themselves to particular blacksmiths. In the context of the relationship between farmers and the blacksmiths, payment and labour expended were not direct, immediate nor exactly standardized. It needed an organization that was cohesive like the guild of the blacksmiths to maintain the balance necessary for harmonious coexistence.

The blacksmiths' guild took care of the situation by various means. The guild wielded a very powerful instrument of social control in 19th-century Zamfara. This was affected through social ostracism. Here they employed the services of the *mai zari*, a musician who was the smiths' professional praise singer, to continuously harass the culprit. The smiths employed the *mai zari* to punish those who refused to settle their debts to the smiths, those who damaged his property by allowing animals to graze on his farms, or those who broke the sexual moral code. The punishment took one of two forms depending on the severity of the offence. The first called *Rawasa* consisted of a rude song chanted in the streets. This was continued until the culprit recanted and observed the socially prescribed atonement. The other sanction was carried out by means of witchcraft. This was reserved for serious offences. In the former case, the culprit could be harassed to the extent of leaving the town. In the latter case, it was believed that the culprit either died, went mad, or lost his manhood qualities. With this powerful instrument of control at their disposal, it is therefore clear that the smiths' guild could not only represent the interest of its members to the state, it could protect the same against offending members of the society.

One other method employed by the blacksmiths in affecting conformity was by embarking on "work-to-rule." The *Mazuga*, on receiving reports of interference from political office holders, after verifying the veracity, could call on members to do so. This was the most effective method. When the blacksmiths embarked on "work-to-rule," the society was temporarily brought to a halt. Everybody was thus made to know the offence committed and by whom. This brought considerable shame unto the person or family that had caused the crisis. It was something to be avoided.

The dependence of the farmers on the blacksmiths was varied during the rainy season. Throughout the farming activities of land clearing, sowing, weeding, ridge making and cultivation, the farmers needed different kinds of implements. Hoes were the general implements often mentioned, but even the hoes were varied depending on the activities going on. Thus *galma,* and *kwasa* were employed in *fadama* and tough soil. *Kalma* was used principally in sowing, harvesting and generally on light soil. *Hauya* was the ordinary farming hoe. The implements manufactured by the blacksmiths for the farmers and the society are too numerous to mention. Iman Imoru took the trouble of listing some.[40] Thus, heavy axes, *gatari*, were needed for forest clearing, cutlasses to weed grass, huge hoes, *galma,* to make ridges, and very heavy hoes were needed for ploughing.

In all the above-mentioned cases if the farmers did not require new implements, they needed the old ones to be mended or sharpened. During the dry season, there was a lull in the demands of the farmers on the blacksmiths. During such periods the state usually required their services, most especially if there was an impending military campaign. If there was none, there was a ready market, for it was the period when most other crafts were practiced in earnest. Hunting was another preoccupation in the dry season, and so hunters also availed of the services of the blacksmiths during the season.

As regards its relationship with the larger society, not only did the guild succeed in winning recognition from the state, it also won respect from the society during the 19th century. From the state officials through to the farmers and other craftsmen, provision was always made for the blacksmiths throughout the year. Thus, the *Sarki* may send the *Mazuga* gifts of clothing or animals during certain festivals. On festive occasions, the blacksmiths were treated with reverence and deep respect.

Other Metals In 19th-Century Zamfara

In Zamfara as in most other areas of Hausaland, the emergence of the Chieftaincy system, referred to also as the *Sarauta* system, was accompanied by a distinct class culture manifest in all aspects of societal life. Members of the royalty, for example, were not supposed to be engaged in labour, nor were they supposed to partake in commerce, at least not directly. Class distinction was also manifested in comportment and dress. Some other metals in Zamfara were used as part of dressing especially by the female members of the ruling class (whether by marriage or birth), as body adornments to enhance beauty. These included Gold (*Zinare*), Silver (*Azurfa*), Copper (*Gacci*), Bronze (*Tagulla*) and Brass (*Farin Karfe*). These metals were also used to a lesser extent by the male in terms of horse accoutrement. None of the metals so mentioned was produced locally in Zamfara. They were all procured through trade. Silver, Copper and its alloys were obtained through Gwanja. Copper was obtained from the north via Gobir, and Gold was obtained from both the north and Gwanja.

The principal ornaments from these metals were ear rings (*Yan kunne*), armlets (*mundaye*), anklets (*mundayen kafa*), rings (*zobba*), rings on feet fingers (*taka tsara*), and necklaces (*abun wuya*). Brass and bronze were also used to manufacture receptacles, spoons and ladles, decorated in various designs. Bowls, kettles and trays of various sizes were also made out of them.[41] These other metals were also employed in the making of sword hilts.

While whitesmithing may be considered to be relatively simpler than blacksmithing, the range of its products surpassed those of the blacksmiths. The various styles employed needed no less ingenuity. In fact, once a smith was able to cast, the rest depended on his ingenuity. It was a craft whose range of products was limitless. Imagination and aesthetics played a greater role in the whitesmithing industry. This was essentially due to the casting method employed. Since the cast was in clay and had to be broken up, each piece was an original and could not be reproduced. Moreover after the cast various methods were used to enhance the aesthetic quality of the end products.

By c. 1762 district whitesmiths had started emerging in Zamfara. Prior to the division of labour between the blacksmiths and whitesmiths, blacksmiths seemed to have tried their hands whenever these exotic metals became available. With the expansion in the market and subsequent increased demand, the need for specialization arose. This was achieved in one of two ways. The first possibility was for a blacksmith to branch off and concentrate on whitesmithing. The other alternative was for whitesmiths to be brought in from a different society. This type of migration by craftsmen was not uncommon in Hausaland. In fact, most of the whitesmiths in Zamfara were people from Azbin or *Azbinawa*.[42] Nadama estimated that by the 18th century, this system had been firmly established.[43] At that point, they were not independently organized. They were part of the guild of blacksmiths.[44] There was not much specialization in the industry prior to the Jihad.[45]

The Effect of the Jihad on the Metal Industry

There is a consensus of opinion that though the Jihad succeeded immensely in changing the ideology of the areas it affected, the same can not be said of the production base. In the words of Yahaya Abdullahi, "the ideological base for the existence of the state changed without an appreciable change in the form of production organization and relations of production."[46]

Saleh Abubakar also noted that "industrial and agricultural technology remained virtually unchanged."[47]

It seems, however, that there were important organizational changes which had far-reaching impact on the economy. We shall only be concerned with its effect on the metal industry. With the establishment of Sokoto in 1809, after the fall of Alkalawa, there was a substantial influx of immigrants into the town. This process reached it zenith with the movement of the Shehu into the town in 1815 as his permanent abode. The settlement of the Shehu in Sokoto had the effect of confirming the town as the political and cultural capital of the emergent Caliphate. This conferred on Sokoto immense economic advantages. Thus the Shehu was followed by a considerable number of people among whom artisans and craftsmen formed a substantial part. From 1815 onwards, most of the new quarters were settled by craftsmen and traders.[48] Of the twenty new wards, six were for smiths alone.[49] Sokoto, therefore, had a very strong base for the development of a viable economy. The successful outcome of the hostilities between Sokoto and Zamfara in favour of the former in about 1827,[50] also opened up more lands which the supporters of the Jihad proceeded to occupy. This tended to cause "a drift of population eastwards and northwards from earlier centres such as Gwandu and Sifawa."[51] The effects which these developments were to have in Sokoto were replicated in the other urban centres in Zamfara such as Gusau, Kauran Namoda, Gummi and Anka.

Caliph Bello's policy was geared towards economic development, on the one hand, and security on the other. The Caliphate in this respect developed an elaborate army organization which included cavalry and infantry. In addition, it developed an "advanced static defence system."[52] Warriors were settled in the *ribats*, and their sole occupation was the defense of the Caliphate. In the absence of any campaigns they were to keep busy practicing and studying. Slaves were settled along with them to do the farming and other undertakings.

In both economic development and defence, the smiths were to play an indispensable role. The role of the state hinged on the integration of all sectors of the economy into a dynamic whole, involving technology and commerce. In this respect metallurgy took pride of place.

CONCLUSION

The metallurgical industries in Zamfara attained a considerable level of sophistication by the 19th century. The developments in the metallurgical industries were part and parcel of the general socioeconomic and sociopolitical reformations sweeping over Hausaland in general and the Sokoto Caliphate in particular. In smelting considerable finesse was developed in terms of knowledge of the properties of clay and charcoal, the principle of fluxing, and the choice of earth used in furnace construction and its treatment.

The iron industry was depended upon by all the other sectors of the society, and smelters and smiths were the real purveyors of the society's technical competence.

NOTES

1.   Nadama, G., "The Rise and Collapse of a Hausa State: A Social and Political History of Zamfara," Ph.D., A.B.U., p. 384, especially fn. 58.

2.   *Ibid.*, p. 89.

3.   Smaldone, J. P., *Warfare in the Sokoto Caliphate: Historical and Sociological Perspective*, C.U.P. 1977, p. 17.

4.   Hogben, S. J. and Kirk-Greene, A. H. M. *The Emirate of Northern Nigeria*, London, O.U.P. 1968, p. 371.

5.   Nadama, G., *op. cit.*, p. 1.

6.   *Ibid.* p. 2.

7.   *Ibid.*

8.   *Ibid.*

9.   Arnett, E. J., *The Rise of the Sokoto Fulani (Being a paraphrase and in some parts a translation of Infakui Maisuri of Sultan Mohammed Bello)*, 1922, p. 77.

10.   Abdullahi, Y. A., "Social Relations of Production in Peasant Agriculture: A Case Study of Simple Commodity Production in Kauran Namoda Area of Sokoto State," Ph.D., A.B.U., 1983, p. 64.

11.   Al-Fasi, Ibn Mohammed Al-Wezaz (Lee Africanus), *The History and Description of Africa and The Notable Things Therein Contained* (translated by John Fory, 1600, edited with an introduction and notes by Dr. Robert Brown), Vol. III.

12.   Nadama, G., *op. cit.*, pp. 127-129.

13.   Smith, M. G., "The Hausa System of Social Status."

14.   Interview with Musa Mazuga, Talatar Mafara, at Muzuga on the 14th September, 1982.

15.   That was the response of most people interviewed.

16.   Smith, M. G., *op. cit.*

17.   *Ibid.*

18.   Interview with Mal. Garba Godai at Gummi, 2nd September 1982.

19.   Jones, B. "The Sedimentary Rocks of Sokoto Province." *Geological Survey Bulletin,* 180, 1948, pp. 56-61.

20.   *Ibid.*

21.   *Ibid.*

22.   *Ibid.*

23.   *Ibid.*

24.   Anonymous, Bulletin, Department of Geology, A.B.U. Zaria, Nigeria, 2, 1, 799, p. 325.

25.   *Ibid.*, p. 349.

26.   Interview with Saraki Ango and others at Maradun, 24th September, 1982.

27.   Sundry Field Interviews.

28.   Interview with Mazuga Muhammadu at Tureta, 31st May, 1985.

29.   Adamu, M., "A Survey of The External Commerce of the Hausa People From About A.D. 1500 to Date," p. 4 paper presented at the Kano Seminar on the Economic History of the West African Savanna. 1976.

30.   Mosely, K. P., "Caraveland Caravan: Long Distance Trade and West African Development in the Pre-Colonial Period," Kano Seminar.

31.   Adamu, M., *op. cit.*

32.   The zango (sites) mentioned are Danfawa (Zurmi), Walu (Kauran Namoda), Dutsin Disa (Tureta), Birnin Magaji (Gummi), Zangon Gado dan kigo (Bukwium), Zangon Sule (Denge), Zangon Sule (Raba), Damaga (Sokoto), Ganajaye (Binji), Dan Kell (Wamako), Illela, Lugu and Gandaba (Yabo), Salame, Dan fashi, Kaurar Wanka, Zoma (Gummi), Dajin Yandu, Wawan ice, Tabkin danko.

33.   Interview with Mazuga, Anka, held on the 29th September 1982.

34.   Interview with Saraki, Ango, at Maradun, 24th September 1982.

35.   It is variously estimated as up to the chin of, or the armpit of, a man, or even the height of a man.

36.   Interview with Sarkin Makera Mazuga, Talatar Mafara, in Maradun, 24th September, 1982.

37.   Krieger, K., "Notes on the Iron Production of the Hausa" (trans G. Seidensficker). *Zeitschrify Fur Ethnologie*, 85, 1963, pp. 318-331.

38.   Enchard, N., "Notes sur les Forgerans De Lader," *Journal de societe des Africanistes*, 35, 2, 1965, pp. 357-372.

39.   Krieger, K., *op. cit.*

40.   Fergurson, D. E., "Nineteenth Century Hausaland: Being a Description by Iman Imoru Of the Land, Economy and Society of His People," Ph.D., U.C.L.A., 1973, pp. 327-330.

41.   These were often imported ready made especially from Nupeland.

42.   Nadama, G., *op. cit.*, pp. 127 and 198.

43.   *Ibid.*

44.   Group interview, Gummi, 2nd September 1982.

45.   *Ibid.*

46.   Abdullahi, Y. A., *op. cit.*, p. 71.

47.   Abubakar, S., "Birnin Shehu, The City of Sokoto: A Social and Economic History, C.1809-1903," Ph.D., A.B.U., 1982, p. 136.

48.   *Ibid.*, p. 40.

49.   *Ibid.* p. 69.

50.   Nadama, G., *op. cit.*, p. 475.

51.   Abubakar, S., *op. cit.*, p. 34.

52.   Smaldone, J. P., *op. cit.*, p. 38.

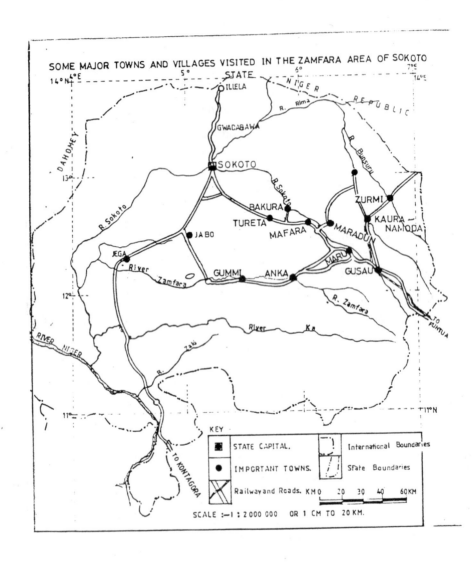

SOME MAJOR TOWNS AND VILLAGES VISITED IN
THE ZAMFARA AREA OF SOKOTO

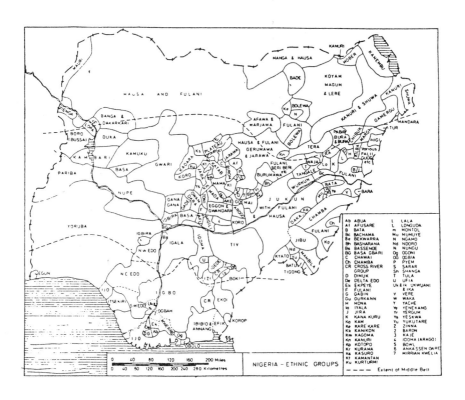

ETHNIC GROUPS

NIGERIA

AFRICA

Chapter VI

Gold Mining in Pre-Colonial Zimbabwe

L. R. Molomo

One of the important functions of historiography is to explain the past; that is, to furnish answers to the questions as to what happened in the past, how and why it happened in the way it did, and at the time it happened. This is true of all historiography: but it is particularly so with respect to African historiography. Africa is one of those regions caught up in a vicious and debilitating phenomenon of underdevelopment. There is an obvious need for historians of Africa to dig into the past to find the roots of that phenomenon. It is plausible to suggest that once the peoples of Africa can understand how they came to be trapped in the vicious circle of underdevelopment, they will most likely discover ways of coming to grips with the problem. This is neither the 'search for African initiative and agency'[1] nor the reiteration of 'the obvious point that Africans were instrumental in making their own history'[2] of which Africanist historians have been accused. Nor is it what Hopkins has called the 'myth of Merrie Africa'[3] against which he sternly warns Africanist historians. These views have been jubilantly welcomed by I. R. Phimister who celebrates the condemnation of the mystifying notion of 'Merrie Africa' in the following words:

> In particular, Africanist interpretations of that nature provide a conceptual tool which is of limited value for probing the origins

of Africa's technological gap and one which rarely even begins to grapple with the historical roots of Africa's present under-development.[4]

This chapter examines the development of gold mining technology in pre-colonial Zimbabwe. Gold mining, like the mining of other metals such as copper and iron, was a long-established industry among the Shona people. But by the end of the 17th century most mines had ceased production "although not worked out."[5] This rather abrupt abandonment of an industry that had deep historical roots in the economy of the Shona is perplexing. The consequences were disastrous. The Shona not only lost lucrative trade in gold but, perhaps even more disastrous, a process of technological development which had begun more than a thousand years before was brought to a halt. In this chapter the author attempts to analyze *the development potential* of pre-colonial gold mining technology in Zimbabwe and to estimate the extent to which the operation of external factors upon the Shona pre-colonial society may have been instrumental to its arrest.

Scholarly attempts have been made in the past to explain this apparently incomprehensible event. In this regard we must single out the monumental work of Roger Summers. In his pioneering work *Ancient Mining in Rhodesia And Adjacent Areas* Summers has described, inter alia, the Shona prospecting and mining methods as well as the difficulties with which the miners had to contend. He further made a laudable if rather hazardous effort to estimate levels of gold output at various stages in the history of ancient mining in Zimbabwe. He illustrated the production levels with a hypothetical histogram covering the period c. 600-1883 A.D. The histogram shows a rather steep fall in production after the 14th century. This would suggest that the decline in Shona gold mining began well before the arrival of the Portuguese in 1506. But there are indications that gold mining in pre-colonial Zimbabwe did not cease until well after the 17th century or later. This issue is discussed in greater detail below. For the moment suffice it to say that Roger Summers has made a sterling contribution to the understanding of this important area of the history of Zimbabwe.

Roger Summers was a professional archaeologist and his work laid a foundation for other scholars working on the history of Zimbabwe. The most modern historian who built on that foundation is I. R. Phimister. Phimister has not only done a D. Phil. thesis on mining in Zimbabwe, but he has also published scholarly articles in reputable journals and periodicals on the same subject. In one of his recent articles Phimister focused on pre-colonial gold mining in Zimbabwe. In this article he raised the question of the date when gold mining in pre-colonial Zimbabwe came to an end and the possible reasons for it. After listing "a variety of reasons such as internal tensions, wars, external intervention and shifting trade routes" which he says were not of long-term significance, Phimister writes: "Undoubtedly the central reason for the decline in reef mining was that the mines were becoming worked out"[6] by the 17th century. He then gives as second in importance the low prices that Zimbabwean gold mining fetched as a result of unequal exchange. Prof. Phimister's conclusion, then, is that pre-colonial gold mining in Zimbabwe came to an end in about the 17th century because the mines were exhausted and because gold fetched low prices on the international market. In view of the important implications of this conclusion for development theory, frequent reference to it will be made below. In the meantime we proceed to analyze the dynamics of the pre-colonial gold mining in Zimbabwe.

Mining is the process of extracting useful minerals from the earth's crust, through surface or underground workings. Knowledge of gold and goldsmithing appears to have originated in Nubia in the area of modern Sudan in about 6000 B.C.[7] Gold was found deposited in the riverbeds as placer gold. From the Nile Valley goldsmithing apparently travelled to the Middle East, the Far East and Western Europe and the Iberian Peninsula. Evidence for West Africa suggests that gold was mined as early as the 4th century, though production did not peak until after the 7th century.[8] Reference to trade in gold in East Africa occurs for the first time in the 10th century:[9] but there are indications that the gold may have originated in Zimbabwe. In Zimbabwe itself there is no clear evidence and agreed date as to when gold mining began. Roger Summers, relying partly on his own speculations and partly on records of Arab geographers, suggests 600 A.D. as a possible date. But Phimister, citing his own 'arguments' and "current research"[10] would have the 9th or 10th century as the probable date for the beginning of the 'ancients' gold reef mining. Meanwhile Professors Roland

Oliver and Brian Fagan have drawn attention to two of the Ziwa sites which they say have been associated with ancient gold workings. Of the four radiocarbon dates that have been obtained from these sites one is in the 4th century while the remaining three lie in the 9th and 10th centuries.[11] The dates appear to support Phimister's argument for the ninth and tenth centuries. A further support for the tenth century comes from Elspeth Jack who talks of Sotho gold miners who lived in Zimbabwe "as early as A.D. 900."[12] It should be noted, however, that Phimister's dates refer to 'gold reef mining'; not to alluvial or surface mining which may well have begun many centuries before. The picture that seems to emerge is one of gold mining at alluvial mining stage from about 300 A.D. and reaching underground gold reef mining stage by the 9th or 10th century.

How and why was the attention of the 'ancients' attracted to the gold metal? Gold is found glittering in riverbeds, below gravels, in depressions above bedrocks and in rock outcrops. It is not clear whether by the time of their arrival in Zimbabwe, the Early Iron Age migrants had the knowledge of gold and goldsmithing. It is certain, however, that they mined iron and forged iron tools and weapons. Their mining experience, it is plausible to suggest, would have stood them in good stead not only in detecting alluvial gold but also in goldsmithing. Once they realized that gold was malleable and that its attractive glitter was everlasting, the 'ancients' would almost certainly have made a conscious effort to find it.

Be that as it may, it would appear that gold was sought for its ornamental and ceremonial value. Gold objects or ornaments were used on ceremonial or state occasions. Gold bracelets and anklets were also worn for personal adornment. But more importantly, it would appear, gold may have played a key role in the funerary ritual. Archaeologists have excavated large quantities of gold objects ranging from gold beads, earrings, wound wire bracelets and anklets to gilded staffs or mace-heads and headrests. These materials have been excavated from many burial and ancient storage places in Zimbabwe.[13] It is evident that gold may have played an important role in the cultural and spiritual life of pre-colonial societies in Zimbabwe. This may have been the real reason why gold was mined in the first place. Trade which developed on the east coast, first with the Muslims and later with the

Portuguese, was seen as no more than a useful source of luxury possessions which enhanced one's social status. This would seem to support Phimister's insistence that gold mining in pre-colonial Shona society was "a marginal and discretionary activity,"[14] always subordinate to agriculture. It never became the backbone of the Shona economy. Trade was doubtless an important adjunct of the Shona economy, but it was always subordinate to agriculture and security considerations. One of the reasons why gold production declined in the 17th century was that the Portuguese traders were destabilizing and endangering the security of the traditional society, that is, the social, cultural, spiritual, economic and political life of the Shona polities.

The methods of finding the gold would, of course, have developed from their native familiarity with their inorganic environment and ecosystem. Once they have known what gold looked like, it was easy for them to identify it in streambeds and rock outcrops. With time the 'ancients' came to know the characteristic appearance of the gold-bearing rock and the typical features of land forms which were former streambeds and therefore likely depositories of placer gold. Roger Summers has claimed, with respect to Zimbabwe, that different kinds of rocks yielded types of soil that tended to 'support' specific forms of vegetation.[15] If the ancient miners discovered this relationship, then trees may have been important guides for the pre-colonial prospectors. Also, antheaps may have been important pointers in the 'ancients' repertoire of prospecting methods. Ants (termites) bring up and deposit soil in the form of anthills. By panning samples of this soil, one could easily establish the presence or absence of gold particles. Initially the prospecting procedure would have involved only detection and collection of alluvial and eluvial gold. But as these surface deposits were being exhausted it became necessary to develop new ways of identifying placer gold in former streambeds now lying beneath layers of overburdens. Detection of overburdened deposits and reef outcrops required no more than keen observation and understanding of the action of running water and wind on surface materials. The 'ancients' were keen observers and were quick to detect casual relationships between natural elements in their ecosystems. Prospecting for underground quartz reefs may have posed a difficult problem for the ancient miners; but the problem could not have been insoluble. All the ancient miners needed to do was to scan a locality for auriferous rocks. Plentiful surface occurrence of such rocks was seen as evidence of the presence of underground quartz reefs. There is no reason to suppose that the 'ancients' acted differently. This method of prospecting did not require any sophisticated technology. It has remained basically the same in our own time. In pre-colonial Zimbabwe the country's physical features may have made the method even more important than is often realized. Most of the rivers, particularly those flowing north into the Zambezi, tumble down from the main watershed, what D.S.G. Thomas has called the Central Axis,[16] with great velocity and rapids. This means that the rivers deposited only slight amounts of gold in alluvium. The obvious result would have been to reduce the importance of placer mining and of prospecting for alluvial gold in favor of the search for quartz reefs and gold reef mining. It has, indeed, been estimated that ninety-five percent of the ancient gold workings were on quartz reef.[17] Prospecting for alluvial gold, therefore, may have been important in the initial stages, but as the industry gradually passed onto quartz reef mining, different methods became necessary. This represented an advance in the development of prospecting techniques.

The advances in prospecting techniques were matched by those made in the extraction and procurement of gold. Development was necessarily slow and evolutionary. In the period before the 10th or 11th centuries, surface mining would have predominated. The miners simply collected sands/gravels from streambeds and from other secondary deposits and panned them for gold particles. The 'ancients' did not, of course, know the exact specific gravity of pure gold. But they knew that it was heavier than most of the soils and gravels with which it was mixed and that by panning it they could separate it from these materials. Placer gold was also collected from the banks of rivers and terraced land surfaces. Sand and gravel from these were shovelled into pans which were then filled up with water. By simply twisting and tilting the pan, the lighter gravels were thrown off while the gold grains settled down and remained in the pan. This stage of surface mining was probably the easiest in terms of prospecting skills and physical exertion in the actual extraction of gold. But placer gold sometimes occurred in low-lying terrains which were once river-courses and were then filled with thick overburden. The miners had to remove the overburden by strip methods in order to reach the deposits. Finally, there were cases where gold was found in surface outcrops which dipped for not more than fifteen meters. But these were often under hard rocky overburdens. The ore in such cases had to be extracted by open-cut methods. But as the reef outcrop dipped almost vertically the open-cut was converted into a step. Steps were made to facilitate up-and-down movement by the miners. The steps were also used to position the 'human chain'[18] that hauled the ore to the surface. One person was stationed on each of the steps. The bottom-most received the ore from the miners. It was then passed from hand to hand until it reached the surface. Where there were no surface outcrops shafts were sunk. The depth of the shafts was determined by geological factors; but there are indications that the shafts could be as deep as 45 meters. Once the shaft reached the gold vein, the latter was exploited in all directions until it was exhausted.

The most remarkable achievement of the pre-colonial miners in Zimbabwe was, perhaps, their assemblage of mining tools. At the alluvial stage of mining the main tools were, of course, the panning tray and the scooping bowls. Both were made from wood. But as the industry moved from alluvial to reef gold mining more appropriate tools had to be found or fashioned. The first to be forged was the iron hoe, initially fashioned for agricultural work. In mining the hoe was used to break, dig or draw the soil. It was also used to gather together broken rocks and to draw them into containers. The 'ancients' also used stone boulders and hammerstones to pound the quartz, though continuous use soon smoothed the subject rock and rendered the hammerstone ineffective.[19] By far the commonest tool was the iron-gad, which was used as a chisel to split the rock. Most gads were pointed on both ends so that one end could be stuck through and held in a knobbed wooden handle. Where the rocks were too hard to be cracked by any of the above methods fire-setting was used. This involved heating the rock and suddenly cooling it off by pouring cold water on it. This 'quenching' of the hot rock eventually shattered it to pieces. The broken pieces were then taken in wooden bowls for crushing and washing. It is clear from this that Zimbabwean pre-colonial miners had developed fairly sophisticated technology. There is evidence that by the 19th century Zimbabweans knew and could manufacture gunpowder. It is plausible to suggest that had there been no European intrusion, they would have discovered its use in the mining industry.

The Zimbabwean pre-colonial performance is even more impressive when seen against the background of the problems with which they had to contend. The 'ancients' had to contend with problems of ventilation, flooding, lighting, hauling and hoisting. Ventilation was doubtless one of the most serious problems that the pre-colonial miners encountered. But even here the 'ancients' showed innovative resourcefulness. The first obvious way of overcoming this problem would have been the use of bellows. Zimbabwean pre-colonial gold miners were accomplished iron-smelters and would certainly have been familiar with the bellows. There is every reason to suppose that they applied "the bellows principle" in the provision of air to the gold workings. The suggestion that they used wetted linen cloth, substituting "a brayed skin" later may need to be reviewed.[20] Manually operated bellows forced currents of air through the workings. Another method that was used by the 'ancients' to ventilate their workings was a multi-shaft system. Two or more additional parallel shafts were sunk and connected by cross-cuts to facilitate the circulation of air. The 'ancients' also applied the 'convection' principle to ventilate the gold mines. Fire was lit at the bottom of the shaft. The hotter and therefore lighter foul air would rise up while the cooler, fresher and heavier air, was drawn down to take its place. These methods did not finally remove the problem of ventilation that Zimbabwean gold miners faced. The deeper the shafts went the more acute the problem became.

The second major problem that faced Zimbabwean pre-colonial gold miners was flooding and underground water. The mine drainage problem was undoubtedly one of the causes of premature closing of the mines. Pumping skills, and equipment were the basic constraints for the ancient miners. But this alone did not bring the gold mining industry to a halt. The miners took advantage of the seasonal character of the ground-water level. Mining was done during the months of August, September and October when there were no rains and the water table was low. But however low the water table might be, the basic problem remained. The mines could not go below the water table. "Once that point was reached, work was abandoned."[21] "Sometimes it happens that such a rush of water bursts into the mine that it is flooded and it is impossible to extract the mataca."[22] Although the 'ancients' would have attempted to remove mine water with buckets, more effective methods were obviously needed.

Most of Zimbabwe's pre-colonial gold workings were on quartz reef. Roger Summers' estimate is ninety-five percent.[23] Where the veins dipped steeply the common method of quarrying the gold-bearing ore was open stoping. Slanting pits with steps to facilitate up-and-down movement were dug. The majority of the stopes were about fifteen meters long, but there were some that were over two hundred forty meters. In spite of the drainage problems noted earlier on, the majority of the ancient workings were between twelve and fifteen meters deep. Here again there were exceptions to the norm: a significant number of the stopes measured forty-five meters deep. Although by our 20th-century standards these workings may be classified as 'shallow,' there can be no doubt that for the 16th and 17th centuries they represented an impressive achievement. In some areas of Zimbabwe the gold quartz reefs occurred in relatively flat deposits. The reefs here are more or less horizontal. In such cases near-vertical shafts were sunk until the vein was reached. The gold-bearing ore was quarried in all directions, using underground stoping. The ore was broken off with stone-hammers and iron-gads. Rocks which were too hard to break off were cracked by fire and then split with iron-gads. The ore was then hauled to the surface for milling and washing. The question of haulage or hoisting of ore to the surface was not as insurmountable as the problem of removing water from the mines. The miners extracted the ore from the veins and handed it to the haulers who stood in a 'chain' and passed the ore from hand to hand in wooden bowls until it reached the surface. Leather bags may also have been used to haul the ore to the surface. These had straps which were slung onto the forehead while the miner crawled out of a shaft. Be that as it may, once the ore had reached the surface it was carried to the milling sites where it was crushed into fine powder. The ore was probably placed on the grinding stones or mortars prepared for the purpose and then pounded with hand stone-hammers or boulders. The powdered ore was then washed in order to separate gold particles from the gangue. It was panned in wooden trays. But there are evidences that the 'ancients' may have used stone riffle-plates with dolly-holes.[24] The dolly-holes had rough bottoms to trap gold particles while undesired gangue was washed away. It would appear that, unlike the West African ancient miners, pre-colonial Zimbabwean gold miners did not use mercury to concentrate the gold. This would seem to point to the indigenous origins of Zimbabwean gold mining industry. Claims by Roger Summers that Zimbabwean gold mining had Indian links are not borne out by the evidence.[25] Mercury was widely known and used by Indian miners in the amalgamation process, but so far no evidence has been found of amalgamation techniques in Zimbabwe's pre-colonial mining. Moreover, if gold mining in Zimbabwe began in the 4th century A.D., as it has been suggested (above), then one can hardly derive its origins from India. Both the two stages of placer mining, namely, alluvial and overburdened deposits, and the quartz reef mining depended on panning and sluicing for the procurement of gold.

The gold mining industry was internally well organized and tightly controlled. Mining operations were undertaken by the entire village or community under the direction and strict supervision of the chief or village headman.[26] The chief could, for instance, order the closure of a rich mine if in his opinion continued operations appeared to endanger the security of the polity or state. The final stage of the mining operations, goldsmithing, was the work of the specialists. Portuguese sources, quoted by Theal, seem to suggest *individual diggers* sold gold to buyers directly.

This is manifestly inimical to the communal approach of the Shona mining practice. We have just noted that each mining operation was undertaken by a whole community under the direction and control of the chief or headman. It was to this chief or headman that the Munhumutapa sent "one or two cows" and received gold in return. Phimister's interpretation that "ordinary laborers seemingly retained the right to sell the products of their labor directly to traders"[27] needs to be reviewed. Goldsmiths worked under the close supervision and control of the chief or headman, and the worked gold was probably stored at his residence. After their occupation of Zimbabwe in 1890 the British ransacked and looted all the great ruins, once royal residences, and removed all the gold they could find. In 1902 R. N. Hall estimated that between 1895 and 1902 about 2,000 ounces (about 62,000 grammes) of gold was removed from Great Zimbabwe alone. Nine years later, on January 30, 1911, he told the Rhodesian Scientific Association that while he was a Curator of Zimbabwe Ruins (1902-1904) he found "about four thousand pounds worth of gold."[28] Great Zimbabwe was not the only ruin that had been an important residence of the ancient rulers. Others were Khami, Dhlodhlo, Natalie, etc. These were similarly ransacked and plundered during the 1895-1902 treasure hunt. The presence of these royal centres of large stocks of gold testifies to the fact that gold mining was not only a cooperative undertaking but also that the gold was stored at the official residences of the leaders. Individual private sales in the circumstances would have been illegal and isolated occurrences.

We have, so far, discussed the various aspects of the gold mining industry in pre-colonial Zimbabwe. We have seen the impressive results that the ancient miners were able to achieve by the 16th century. But we have also seen the severe limitations with which the 'ancients' had to contend. In this second part of our discussion we take up the issue of the decline and final collapse of the 'traditional' gold mining in Zimbabwe. The question to be addressed in this connection is: Why did the gold mining industry decline after the 16th century? According to Roger Summers' hypothetical histogram[29] production was at its peak between the 10th and 13th centuries. Phimister has put forward an alternative hypothetical histogram according to which the industry peaked between the 12th and the mid-15th centuries.[30] By the 17th century, on the authority of the Gold Trade table,[31] production had fallen to mere a 5,000 ozs. (about 156,000 grammes) per annum. Roger Summers' statement that "by the seventeenth century many of the mines were becoming worked out, at any rate so far as 'ancient' payability was concerned" is hardly helpful. His observation is contradictory. Roger Summers further states that tenor was falling below 16 to 20 dwt per ton. What was happening? Was gold in the mines exhausted? If so, how come "many mines were actually closed down before being worked out"?[32] After the British had occupied Zimbabwe in 1890 they found extremely rich gold deposits (later Mickey Mine and Lonely Mine) near ancient workings. It is not to be supposed that these deposits escaped "the sharp eye" of the 'ancients.' As will be demonstrated later on, the authorities may have refrained from mining the gold for security reasons. In about 1630 a Portuguese official at Sena, Albarado, was told by a local chieftain that there was much gold in Monomotapa and that nuggets had been found up to 3,000 mithqals in weight; "but as soon as a mine of substance was found, the Emperor ordered it to be closed under penalty of death, lest it be 'the mother' of the gold."[33] Soon after 1505 a Portuguese official at Sofala stated that "*in time of peace* Sofala exported a million, or even 1,300,000 mithqals of gold a year."[34] What can we make of this situation? It would appear that until well into the 17th century the Shona miners could still have produced at optimum levels, but that considerations of political security operated against the attainment of these levels. It was not so much the fall in the ancient mines' payability as the fewness of the mines that adequately explains the low export figures. Malyn D. D. Newitt has drawn attention to the fact that early in the 17th century the fairs became centers of military activity during the wars, and this probably had the effect of "driving away the native inhabitants and putting an end to the local panning of gold on which the fairs depended for their prosperity."[35] There are strong indications that the Portuguese trading activities were accompanied by unacceptable degrees of violence, and that the Shona leaders were concerned as to the effect this might have on the political stability of their society. No sooner had they established the linkage between the gold trading activities and the violence than they moved to withdraw supplies. The conviction on the linkage was brutally expressed in an incident in which a Quiteve was actually dethroned and killed by his chiefs because it was believed that he was going to open the mines and allow the Portuguese to work them.[36] It is evident from what has been said in this paragraph that the contraction of output in gold mining was owing more to political destabilization than to any other single factor.

From the time of the arrival of the Portuguese in the Zambezi Valley in 1505 Zambabwe became a scene of unprecedented violence. The first and, perhaps, the most shocking and traumatic to the Africans, at any rate, was the massacre of Muslim traders in about 1570. The incident was vividly described by Father Dom Goncalo as follows:[37]

These men (the Muslim traders) were condemned to death by strange inventions. Some were impaled alive; some were tied to the tops of trees, forcibly brought together, and set free, by which means they were torn asunder; others were opened up the back with hatchets; some were killed by mortars, in order to strike terror into the natives; and others were delivered to the soldiers, who wreaked their wrath upon them with arquebusses.

In its sadism as in its brutality, the massacre could hardly be equalled in the history of Christendom. While it may have provided the Portuguese with an opportunity to wreak their vengeance upon the Muslim traders, it did not "strike terror into the natives." On the contrary, it left them with a deep sense of disappointment and apprehension and it opened their eyes, as it were, as to what they might expect themselves. The Zimbabweans did not have to wait for long before their fears were confirmed. The last two decades of the 16th century witnessed escalation of violence on a scale hitherto unknown in the Zimbabwean society. A series of Portuguese-inspired rebellions by vassal chiefs against their authority forced the Munhumutapas to appeal to the Portuguese captains for military aid. Such assistance often cost the Shona monarch not only their power and prestige, but also the country's basic resources. In 1607 Munhumutapa Gatsi Resere had to cede all the mines in his kingdom in return for aid. The implications of the cession for the gold mining industry are considered below. Twenty-two years later, in 1629, Munhumutapa Mavura was forced to cede the whole of his kingdom to the Portuguese crown.

These political disasters, at any rate for the Zimbabweans, have an important bearing on our question: why did output in gold production decline after the 16th century? At the beginning of the second part of this chapter, it will be recalled, we suggested that political rather than technological or trade factors lay at the root of the decline. The dates of the political disasters referred to in the preceding paragraph speak for themselves. The years 1590 to 1630 were years of unprecedented violence, involving Africans but instigated and exploited by the Portuguese. As early as 1512 they had begun to penetrate the interior from the east coast and to settle in the Shona country. Although gold was their main quest the Portuguese, most of them criminals of gigantic proportions, were prepared to grab at wealth in any form. In their search for wealth they meddled in the local politics; fueling and exploiting traditional feuds among local rulers, instigating rebellions by vassal chiefs against central authority and furnishing mercenary services in exchange for mineral resources or land. Some of the Portuguese married into African families, acquired land and built retinues of slave armies around themselves. In extreme cases, the Portuguese deposed or killed recalcitrant chiefs and replaced them with their own nominees. Newitt has recorded an incident which occurred in the 1630s in which the Portuguese "dethroned and killed one Chicanga, placed their own nominee on the throne...."[38] This state of lawlessness and political instability greatly disturbed the Shona leaders and had its sequel in war in which they sought to expel the Portuguese from the Zambezi Valley. In 1693 war drove the Portuguese from the Highveld to the very gates of Tete. Their helpless situation was saved by the untimely death of Changamire Dombo and the arrival of reinforcements from Lisbon.

It is evident from what has been said in the preceding paragraphs that contraction in output in the Zimbabwean gold mining occurred in a period of extreme political instability and social insecurity. It could not have been otherwise. Given the conditions of war and rapine such as obtained in Zimbabwe during the period under consideration, no amount of technological advancement and favorable terms of trade could have staved off the decline. Nor is it being denied that the low price of gold on the international market may have been a disincentive. Ventilation, lighting, drainage and haulage were never permanent disabling factors in the history of mining.[39] Why did they become insurmountable in Zimbabwe's gold mining? The truth of the matter seems to be that they were not insurmountable. Zimbabwean miners were never given the chance and opportunity to experiment with their natural environment in peace and independence - a necessary and indispensable prerequisite for the discovery of the natural laws on which the environment operates.[40] External intrusion in the form of first Muslim traders and later the Portuguese appears to have brought no significant novel ideas to the Zimbabwean mining scene.[41] Indeed, it appears to have been more parasitic and exploitative than enriching to the mining industry. But when it exploded into violence in the years 1590-1695 it became fatally destructive. The Portuguese takeover of the mines in 1607 and the assumption of political suzerainty by the Portuguese crown twenty-two years later not only reduced the Karanga monarchy to a figment of its former glory, but it also left the Karanga people with a deep sense of revulsion. Economically it meant that the Portuguese had to organize the mining of gold and provide security for the miners and the merchants. Zimbabweans, of course, neither trusted the Portuguese nor accepted their overlordship. The catalogue of their acts of violence, including the grisly massacre of Muslim traders in 1570, was too long and vivid in their minds to forget and too gruesome to forgive. But perhaps the most unforgivable crime that the Portuguese committed was the seizure of African land transforming it into private estates known as prazos, vulgarization of African women and the seizure and sale of children into slavery. As Newitt aptly put it "such sweeping inroads into the independence of the old Kingdom explains Kaparidze's success in uniting African feeling against the Portuguese.[42] One response of Africans to the Portuguese pressure was to retreat farther into the interior away from Portuguese control and influence. The result of all this was a fall in production levels, not because the miners could not overcome technological obstacles but because military, political and social conditions were simply not conducive. Social security was at its lowest level. The suggestion that the decline in output levels reflected the miners' response to falling gold prices on the international market has been asserted rather than demonstrated. No facts and figures are available. Nor has it been convincingly established that the Zimbabwean economy in the 16th and 17th centuries was crucially dependent on the gold trade to the extent of being hypersensitive to its fluctuating prices on the international market. Phimister has argued convincingly that gold mining in precolonial Zimbabwe was "a marginal and discretionary activity" subordinate to agriculture and pastoralism.[43]

The conclusion that seems to emerge from our discussion in this chapter is that the central factor in the decline of output in gold production in the 17th century was the poor military and security situation, and that this situation was casually related to the Portuguese presence in the Zambezi Valley. The decline and subsequent cessation of the industry meant stoppage of further development in mining technology. Looked at from this standpoint, the decline can be seen as part of a larger phenomenon that came to be known as a process of underdevelopment. The decline and cessation of the gold mining industry was accompanied by the decline in the development of indigenous skills followed by a dependence on imported technology. But the Portuguese underdevelopment of the Zambezi Valley was unique in that while it thwarted development of local skills it substituted nothing in their place.

Bronze plaque representing Oba Akengboi of late 17th century Benin
in battle dress

Cast Gold Alloy, Ivory Coast-Late 17th century

Lady adorned with Gold jewelry, Senegal 20th century

NOTES

1.　　Alpers, E. A., "Re-thinking African Economic History," in *Ufahamu*, 3, 1973, pp. 97-129.

2.　　*Ibid.*

3.　　Hopkins, A. G., *An Economic History of West Africa*, Longman, 1985, p. 10.

4.　　Phimister, I. R. "Pre-colonial Gold Mining in Southern Zambezia: A Reassessment" in *African Social Research*, No. 21, June 1976, p. 1.

5.　　Summers, R., *Ancient Mining in Rhodesia And Adjacent Areas*, Salisbury, 1969, p. 219.

6.　　Phimister, I. R., pp. 26-27.

7.　　Gregory, C. E., *A Concise History of Mining*, New York, 1980, p. 57.

8.　　Garrad, T. F., "Myth and Metrology: The Early Trans-Saharan Gold Trade" in *Journal of African History*, Vol. 23, 1982, pp. 452-453.

9.　　Ogot, B. A. (ed.), *Zamani: A Survey of East African History*, East African Publishing House, pp. 106-107.

10.　Phimister, I. R. p. 3.

11.　Oliver, R. and Fagan, B., *Africa in the Iron Age*, CUP, p. 102.

12.　Jack, E., *Africa: An Early History*, Harrap, pp. 59-60.

13.　Summers, R., pp. 190-194.

14.　Phimister, I. R., pp. 4-5.

15.　Summers, R., pp. 15-16.

16.　Thomas, D. S. G. "Geomorphic Evolution and River Channel Orientation in North West Zambabwe" in *Geographical Association of Zimbabwe*, (Pamphlet) No. 15, July 1984, p. 12.

17.　Summers, R., p. 14.

18.　Hopkins, A. G., 46; Axelson, E. "Gold Mining in Mashonaland in the Sixteenth and Seventeenth Centuries" in *Optima* Vol. 9, 1958, pp. 169-170; Schofield, J. F. "The Ancient Workings of South East Africa" in *NADA*, Vol. 3, 1925, pp. 8-9.

19.   Summers, R., p. 167.

20.   *Ibid.*, p. 166.

21.   Phimister, I. R., p. 12.

22.   Schofield, J. F., p. 9.

23.   Summers, R., p. 14.

24.   *Ibid.*, p. 176.

25.   *Ibid.*, pp. 116-117.

26.   Axelson, E., pp. 169-170; Schofield, J. E. *op. cit.*, pp. 8-9.

27.   Phimister, I. R., p. 23.

28.   Summers, R., p. 192.

29.   *Ibid.*, p. 195.

30.   Phimister, I. R., p. 16.

31.   Summers, R., p. 195.

32.   *Ibid.*, p. 218.

33.   Axelson, E., p. 169.

34.   *Ibid.*

35.   Newitt, M. D. D., *Portuguese Settlement on the Zambesi*, Longman, 1973, p. 45.

36.   *Ibid.*, p. 73.

37.   Cited in Theal, G. M., *Records of South Eastern Africa*, Vol. III, 1898-1903, 1964 Facsimile Reprint, Struik, Cape Town.

38.   Newitt, M. D. D., p. 46.

39.   Phimister, I. R., p. 27.

40.   Rodney, W., *How Europe Underdeveloped Africa*, Tanzania Publishing House, 1980.

41.   Roger Summers' Indian influence has not been supported by recent archaeological research.

42.   Newitt, M. D. D., p. 52.

43.　　Phimister, pp. 4-5.

Chapter VII

Diamond Mining in Sierra Leone 1930-1980

A. Zack-Williams

In an earlier work I have presented a detailed history of Diamond mining in Sierra Leone.[1] Diamonds were discovered in the country in the 1930s, together with other minerals such as gold, platinum, chromite and iron ore. The diamond discovery was brought to the attention of the Consolidated African Selection Trust (CAST) which by this time was operating diamond mines in the Gold Coast.[2] In 1934 CAST formed a subsidiary company, the Sierra Leone Selection Trust, (SLST) a private company wholly controlled by CAST, to exploit the Sierra Leone deposits. In 1935 the company was given sole rights to mine diamonds in Sierra Leone for a period of 99 years.[3] Since 1935 when actual mining starting, SLST held a monopoly in diamond mining and prospecting until 1956 when this was breached with the establishment of the Alluvial Diamond Mining Scheme (ADMS).[4] Under this scheme private individuals were empowered to carry out alluvial mining activities within certain designated areas. Most of these areas were only marginally diamond ferrous, since good quality diamonds could be found only within SLST's lease.

Two types of mining activity were provided for under the scheme: "individual licensed miners," and "native fires." Under the latter non-Sierra Leonean capital was welcomed provided it was not more than 49% of share capital.[5] The ADMS also legalized the purchase and sale of diamonds by private individuals; these being referred to as alluvial diamond dealers. In 1970, after prolonged negotiation with SLST, the company called the National Diamond Mining Company (NDMC) S. L., was formed. However, SLST retained the remaining 49% of the shares, as well as being responsible for corporate decisions and the day-to-day running of the mines.

Modes of Production Within The Diamond Industry

Within the industry we can identify more than one mode of production, i.e., a capitalist (SLST) mode, and a non-capitalist (ADMS) mode. In what follows, I shall argue that the capitalist mode does not constitute an enclave,[6] i.e., it is not isolated or independent of the non-capitalist mode. Indeed, I shall try to show that the very structure of the non-capitalist mode is determined by the dominant position of the capitalist mode within the industry.

SLST and the Capitalist Mode of Production

Since 1935, SLST has carried out large-scale alluvial diamond mining in two separate areas in Sierra Leone; at Yengema, Kono District; and Togo Field, Lower Bambara Chiefdom, Kenema District. In recent years the company has started mining Kimberlite dykes at Kiodu, Kono District; and Sumbuya, in the Kenema District. The company's operations are much more capital intensive as compared to the ADMS, as can be seen from the figures of estimated capital invested in the ADMS from 1956-77 and figures for SLST (See Tables 1 & 2 below).

SLST prides itself on its systematic mining. The company, unlike miners under the ADMS, has the capacity to employ specialists such as geologists and mining engineers. Much of the prospecting done within the ADMS is based on hunch; SLST's labor force constitutes a proletariat in the true sense of the word. The relation of production is defined by the "free" labor's sale of his labor power in return for wages. The worker has no claim to the final product of his labor, since this is the property of the shareholders of SLST. By contrast, as we shall see presently, under the ADMS, the mode of remuneration does not constitute "wages," but assumes the form of "percentage sharing," with the laborer having a legal claim to the final product. Thus under the "SLST Mode," there is a class of laborers divorced from the means of production, and from whom surplus value is expropriated by a class of non-producers, the shareholders of SLST. We shall see below that this kind of distinction (proletariat divorced from the means of production, and capitalists, owners of the means of production) is very difficult to draw for production under the ADMS.

Furthermore, not only does surplus labor assume the form of surplus value under SLST production, but it is appropriated through commodity exchange. This presupposes the existence of laborers who are willing and "free" to sell their labor power. By contrast, we shall see that this description does not hold for the "tributors" under the ADMS, most of whom could at best be described as "worker-peasant,"[7] since they migrate between the diamond fields and the peasant family farms. In short, they are not "free" in the sense that they could stay all the year round with their mining employers, since they have commitments to the peasant family farm. Remuneration from mining is not the sole source of the reproduction of labor under the ADMS.

SLST's Contract Mining Scheme 1959-70

In 1959, SLST embarked upon a scheme which came to be known as the "Contract Mining Scheme." Under this scheme, the company selected certain areas within the Yengema lease which were unsuitable for its own large-scale mining operations and gave them to miners from Kono District who would mine them under contract. The reason for this scheme is not quite clear. Van der Laan has described it as "an attempt to improve the methods of the diggers."[8] Much earlier in the same book he suggested yet another reason:

> This plan was to increase the area where Konos could work legally and would end the belief that the interests of the company and of the Kono diggers were always conflicting.[9]

Whilst this latter reason oscillates around what I consider the objective reason, the whole argument of Van der Laan is premised on a naive assumption of a benevolent capitalist. The relevant question is why SLST failed to establish this scheme earlier? Two events occurred in Sierra Leone around this time which must have influenced the company's decision to try and enlist the support of the people of Kono.

The first was the conflict between SLST (through its "parent" company CAST) and De Beers' Diamond Corporation over the price the former was receiving for diamonds sold to the latter.[10] As a result SLST decided in 1960 to bypass the Diamond Corporation and instead sold its diamonds to two independent American dealers. The second was the radical political agitation that took place in Kono District between 1957-1960, through the Kono Progressive Movement, which tried to articulate the interests of those in Kono who could only join mining operations as "tributors," and for whom diamond mining operations brought little, if any reward. The Movement too was supported by chiefs from the non-diamondiferous chiefdoms in Kono District. With these events in mind we can suggest that the creation of the scheme was an attempt to win over allies in Kono for the cause of the company.

However, it is important to note that the Scheme satisfied a number of basic economic problems for the company - i.e., the need to exploit the most marginally diamondifrous areas within the leased area. Most of the sites had already been mined by SLST. Basically the Scheme entailed that certain sections of the lease would be opened for controlled use by local Kono miners. All diamonds won from the sites were to be sold to SLST. The company was not responsible for hiring labor, nor the provision of equipment, although some equipment was rented to contract miners at a specified rate. Contracts were awarded to Paramount Chiefs of chiefdoms in which the contracts were awarded or their nominees. Clearly, the Scheme provided an avenue for traditional rulers to accumulate wealth and consolidate their power. Furthermore, it provided SLST with cheap labor, hence with diamonds from areas which were unprofitable for it to mine.

This latter point can be substantiated by Appendices 1 & 2. The most important columns are: column 4 of Appendix 1 and column 5 of Appendix 2. These show that for all the years for which figures were available SLST winnings were, on the average, higher than those of the contract miners. The tables show that by using contract miners whose "form" of production was essentially the same as the ADMS, and hence non-capitalist, the capitalist institution was able to gain 53,413.5 carats of diamonds without investing any amount of money in the scheme.

ADMS and the Precapitalist Mode of Production Within the Diamond Industry

In order to understand the history and the present structure of the ADMS, I will pose and try to substantiate a number of hypotheses: 1) That at the time of the establishment of the ADMS, SLST had a fairly good knowledge of the diamond deposits of Sierra Leone; 2) That SLST was in a position in 1956 to opt for the richest deposits; 3) That most of the capital that went into the industry was invested in the SLST sector; 4) That the very structure which the ADMS assumes, in particular its pre-capitalist structure, is the result of the dominant position of SLST within the industry, and the need to perpetuate this dominance; and 5) That the ADMS is not only a pre-capitalist mode, but in spite of the development of productive forces, its old relations of production remain relatively unchanged, and that this is the product of SLST's dominance of the industry.

Hypothesis 1

In another work I drew attention to the strenuous efforts made by SLST to find the limits of the Sierra Leone deposit.[11] By the early 1940s the Sierra Leone field was considered one of the richest in the world. By 1953, the two primary sources of the field had been located, one in Koidu Town, and the other in Lower Babara Chiefdom, Kenema District. It should also be noted that throughout this period, the Geological Survey Department continued with its prospecting to which SLST had access.

Hypothesis 2

That SLST opted for and obtained the best deposits can be substantiated empirically. However, this does not mean that by analyzing production figures from both sectors one would expect to find SLST's towering above those of the ADMS. In fact this cannot be the case since the rate of exploitation between the sectors differed. With fixed installations and heavy equipment SLST had to plan its operations to the capacity and lifetime of its installations. Furthermore, as Van Der Laan has argued the profit taxes which companies pay act as an incentive for the companies to smooth out their profits over the years.[12] By contrast the diggers under the ADMS had only a small outlay on capital. Some of their tools were farm equipments which had been converted to mining tools without much expenditure. Furthermore, taxes do not penalize fluctuating operation or profit as it does a large organization such as SLST. Finally, the operation of illicit diamond miners (IDM) has always tended to inflate the total sales figure in ADMS production. Much of this illicit mining takes place within SLST's lease.

Apart from controlling the deposits in the source areas, SLST was also given the option to select a further 200 sq. miles. This option was later reduced to 100 sq. miles after much resentment from the people of Kono District, where SLST had by this time created a *betriebgemeinschaft*. Van der Laan, a writer who could not be described as critical of SLST's role in Sierra Leone, observed:

> Licensed digging in Kono was handicapped in two ways. The 1955 agreement (setting up the ADMS) meant that SLST had first choice in Kono and that *only the poor diamond bearing land would be left for licensed diggers....*[13] (Emphasis mine)

This point is again reiterated by Rosen. He argued that after the ADMS had been introduced, SLST still had control over the original lease in Kono of some 450 sq. miles of some of the richest diamond areas in Africa. Rosen continued:

> ...many of the areas surrendered by the SLST were only marginally diamondiferous and given the heavy capital requirements, unprofitable to mine. Local diggers on the other hand, using simple equipment and carrying limited cost in overheads, could achieve a high return rate in these marginal areas since their investment was primarily labor.[14]

It can also be argued that since SLST was an important export earner for the country, there was the perceived need to preserve the hegemonic position of the company in order to avoid de-stabilizing the country's export trade.

Hypothesis 3

The fact that most of the capital that entered the industry was invested in the SLST capitalist sector should not be in dispute. We have pointed to the fact that SLST is relatively capital intensive in its operation. Most individual miners could not afford the means to purchase large diggers and excavators such as caterpillars and draglines. The fact that SLST received most of the capital that entered the industry enabled it not only to purchase expensive labor-saving machinery and equipment which helped to reduce handling and hence pilfering, but also meant that the company could attract some of the best personnel. Investment figures for SLST are not always available, largely because of the secrecy which surrounds the industry. However, figures are available for the period 1970/71-1976/77, which together with those of the ADMS tend to substantiate our hypothesis.

TABLE 1
Total Yearly Investment for SLST 1970/71-1976/77

Year	Amount (In Le'000)
1970/71	2,100
1971/72	2,400
1972/73	2,000
1973/74	2,330
1974/75	2,540
1975/76	8,500
1976/77	6,700
TOTAL	27,570

Source: Ministry of Development Freetown.

It is almost impossible to know the exact amount of capital that has been invested within the ADMS. However, an attempt has been made[15] using: 1) The American Aid Revolving Loan Fund (AARLF)[16]; 2) Total number of licences issued; and 3) Estimated cost of digging materials. The Fund has provided us with a useful source of information on investment within the ADMS. Since only a proportion of licensed miners patronized this scheme, the data will be supplemented by further estimates.

Between 1960-76 a total of Le723,641,4 sales were made under the Revolving Loan Scheme.[17] During the same period a total of Le274,875,24 loans were offered under the Scheme. This gives a total figure of Le998,516,64. However, this gives us only a partial idea of the total capital that went into this sector. To have a more comprehensive picture we have to take into consideration the amount of capital sunk into the ADMS in the pre-AARLF era, as well as those unsuccessful miners who did not patronize the Fund. In order to achieve this, we need to know the implements that these miners used, and to try to cost them. These included: shovels, pick-axes, jig-sieves and washing pans. Furthermore, we will need to know how many laborers were employed under the scheme since it is labor intensive. Between 1956-71, some 63,128 licenses were issued and some 461,145 laborers were employed within the ADMS.[18] This gives an average for the 15 years of 4,208 licenses issued and 30,743 laborers per year. Now we have to assume that each laborer had a set of mining equipment. This in fact would be an exaggeration of the real situation, since not all tributors used new equipment. However, these figures give us a rough idea of the amount of money invested in the ADMS. Thus if we take the total cost of equipment and multiply it by the number of tributors, we find that over the period 1956-71 some Le1,985,541,2 was spent within the ADMS.

TABLE 2

Total Amount Invested Within ADMS 1956-77 Excluding "Surface Rent."

Total sales by Mines Division	= Le723,641,4
Total Loan under the AARLF	= Le274,875,24
Total expenditure on mining equipment (1956-71)	= Le1,985,541,2
Grand Total	= Le2,984,057,84

Source: Estimated From The Field.

Hypotheses 4 & 5

These constitute the essence of the rest of the paper. The history of the ADMS can be divided into two periods: the period 1956-62; and the period 1963-77. The former can be described, for lack of a better word, as the "honeymoon period." This period was full of all kinds of unrealistic assumptions. Many licensed miners over-estimated the extent of diamond deposits in the country. To a certain extent this unrealistic approach was compounded by the fact that there were reports of diamond discovery over a wide geographical area. The second period marked the emergence of the "supporters," i.e., the emergence of entrepreneurs who were not simply dealers, but were responsible for the maintenance of the tributors. By this time, some degree of realism had entered the industry, and many of the miners who remained in the field realized that in order to mine their plots, they needed more than pick and shove. Since most of the areas declared for ADMS were marginally diamondiferous and in many cases these were areas which had been mined by SLST, it soon became clear that to mine these deeper deposits a greater capital outlay was needed. From now on most licensed miners came to be dependent upon a "sponsor," a "supporter" as they were called, who provided mining materials or capital for various considerations, such as a claim to a proportion of the value of all winnings. It is interesting to note that virtually all the supporters were licensed dealers, the majority of whom were Lebanese.

Early Method of Labor Recruitment

The task of labor recruitment often follows upon the procurement of a mining lease, though the process may be reversed. In the early days, the license-holder or his appointee "the gang-master" toured villages in order to recruit tributors. Once in a village gifts were presented to the village head, and the purpose of the visit then quickly explained. The best time for such a visit was often after the harvest when activities were at their lowest. A "village crier" will be summoned to broadcast the news. The gang-master would then offer to pay all the debts owed by the potential tributors, as well as their local taxes. From now on the security of the laborers was his responsibility. For this reason, the credential of the gang-master or license-holder was thoroughly scrutinized by the village head. In order to make an impression on the villagers, the gang-master was usually accompanied by a native of the village, usually someone who had "made it good" during the illicit diamond mining period (i.e., pre-1956). While the recruitment lasted the accomplice was given every encouragement by the gang-master to dress above the average village inhabitant and to spend freely in their presence, with a view of enticing as many of them as possible.[19]

This method of labour recruitment has its origins in the illicit mining days when there were only gang bosses who were employers. In those days the only items of expenditure were the recruitment and maintenance of the labor force. In most cases the laborers came with their tools, i.e., picks and shovels, though it was not uncommon for these to be provided by the gang-master.

Since this early system of labor recruitment had its origins in the illicit mining period, it is interesting to answer the question, why did recruitment in the illicit period assume this form? To answer this question it is important to point to the hazards involved in illicit diamond mining (IDM) in areas which were always dominated by SLST's security force, as well as the Sierra Leone Policy Force. To entice workers into such confrontational situations with authority called for a good assurance that in the event of being caught, there would be a "patron" who would look after their welfare and those of their family. Clearly, one can argue that had the company's monopoly not been in force, it is quite possible that the emergence of this form of "patron-client" relationship would not have been so prevalent in the industry.

Early Method of Remuneration of Labor

We have shown in the last section that it was quite common for the direct producer to enter the industry with his own tools.[20] This unity of the direct producer to his means of production is an important determinant of pre-capitalist modes of production. If the relations to the means of production render the ADMS as pre-capitalist, the method of remuneration does nothing to challenge this view of the pre-capitalist nature of the ADMS.

The first point to note is that the workers under the ADMS were not wage-laborers, they were tributors.[21] Though the form of remuneration varied, the category "wages" did not yet exist.[22] If the license-holder himself had recruited his labor force, then the winnings were distributed as follows: Two-thirds for the license-holder; and one-third for the tributors. Where a gang-master was involved the distribution was: 60% for the license-holder, 30% went to the tributors and 10% to the gang-master. Superficially, tributing seemed to be a symbiotic relationship, as it provided incentives to both employer and employed. In practice, tributing was highly exploitative with the burden falling upon the tributors.

For example, the winnings were frequently undervalued, since the tributors had no means of establishing their true value. Occasionally, even the gang-master would be ignorant of the market value of winnings. However, very often both gang-master and license-holder colluded and deliberately undervalued the winnings. It is interesting to note that no serious consideration was ever given to the idea of providing a regular wage structure in the industry - outside the SLST sector. One writer has suggested that it is:

> ...probably due to the unwillingness of both the labor force to forego the proceeds of their labor if there are particularly large winnings and the employer to take on an added expenditure burden which might not be recoverable if the production was nil.[23]

Now this explanation is inadequate since it fails to look at the underlying structure of the industry instead it is premised on some short-term outlook on the part of both employee and employer. The relevant question why their counterparts in SLST were willing to forego the proceeds of their labor if and when there were particularly large winnings was never posed. The reason for these differences in the mode of remuneration is due largely to SLST's abundant resources (physical capital, trained personnel as well as rich deposits), which enabled the company to undertake widescale prospecting and forward planning. Thus it was possible to take workers on a wage basis without jeopardizing the company's existence. The type of forward planning was beyond the reach of most miners under the ADMS; who had neither the rich ground to mine nor the wide geographical area to enable forward planning.[24] Thus, the reason for the pre-capitalist nature of remuneration must be sought neither in the psychological make-up of license-holders and tributors, nor simply within the ADMS, instead we must look at the industry as a whole, in particular the dominant position of SLST within the industry.

The Supporter Period 1973-77

The need to mine the deeper deposits meant that sophisticated machinery had to be brought into the ADMS. With the help of their supporters, many alluvial miners invested in such mining equipment as diesel water pumps and in a few cases even in mechanical diggers. In spite of these developments in productive forces and of relations to the means of production, the mode of remuneration of the direct producer still remains largely pre-capitalist. Remuneration in the supporter period assumes the form of percentage sharing of the final sales value.

Unlike the situation with SLST, under the ADMS as it later developed the direct producer continued to lay claim to the final product of his labor until it is finally sold to a dealer. The mode of remuneration under the ADMS, i.e., percentage sharing, is a characteristic of the feudal mode of production; or in the absence of the tie of servitude is more a feature of modes of production where the surplus generated is appropriated collectively, such as under syndication.

The system of percentage sharing varied according to the parties involved. For example, in the case of a tripartite agreement the most common ration was

Tributors	25%
License-holder	25%
Supporter	50%

In the case of quadripartite agreement the distribution was as follows:

Tributors	20%
Mines Manager	10%
Chiefs or Wardens	10%
License-holder/Supporter	60%

The mines manager was in theory an employee of the supporter, who was there to ensure that diamonds were not withheld from the supporter. The Warden was an employee of the Government's Mines Division, responsible for demarcating mining plots.

As we have noted earlier, the ADMS provided for both individual mining and that of "native firms." Through this ordinance a small number of Sierra Leoneans were able to establish thriving mining ventures. The most successful of these was *Leone Trial Mining (LTM)*, wholly owned by Alhaji Abdulai Sesay. The success of this mining venture substantiates our point that the dominance of SLST in the industry, and the need to protect the company's lease area have led to the atomization of the ADMS. Once a relatively large and "rich" lease was provided for alluvial miners it was then possible for forward planning to take place. But the hiving off of such relatively rich deposits posed a serious threat to SLST's hegemony in the industry, as well as its ability to remain the largest earner of foreign exchange.

Alhaji Sesay first entered mining as an alluvial gold miner in 1934, and remained in the industry right through to 1977 except for the period 1941-45 when he was conscripted into the colonial army. In 1956, with the beginning of alluvial diamond mining, he moved into this sector and carried out both river and land mining on his own until 1965, when he had cause to resort to the supporter system due to increased costs of mining operations.

However, in 1970, he decided to abandon the system when it became clear to him that the supporter system was eating deeply into his profit margin. By 1971, he had come to the conclusion that the only form of mining that would be viable in the areas newly released from SLST was large-scale mining. However, in order to achieve this goal it was necessary for these atomized plots to be consolidated - a new kind of "enclosure" in mining. His major task was to try and convince other license-holders that the best way forward was to try and pool resources, in particular to bring all nearby leases under one roof. The task of convincing miners was not an easy one, many license holders were worried about being entangled in new forms of supporter relationship. But it soon became clear to most miners that the only way forward was to accept the idea of land consolidation. By 1973, agreements had been signed with all interested parties, and the Mines Division had given its blessings to the project.[25]

Now with resources (both land and capital) pooled, it was possible for forward planning to be done. Furthermore, it was also possible for mechanical devices such as draglines and caterpillars.

The method of remuneration of labor varied according to whether LTM was working on contract basis with other license-holders or whether it was working on its own. In the case of the former, winnings were distributed equally between LTM, tributors, and the landowner who had contracted for his plot to be mined. When no contracting party was involved, one-third of winnings went to the tributors, and two-thirds to LTM. The mines manager shared with the tributors.

However, there was one additional feature: LTM paid its machine operators wages. As far as I could establish only one other native firm paid its operators wages. Another miner from Yomandu, who was building a dam on the Bafi River, rented a dragline excavator from the Ministry of Works and as such was not responsible for the wages of the operators.

The question we must now pose is why LTM was able to enter into a wage relation with its operators unlike most native firms. The answer could be found in the fact that LTM was able to employ large mechanical excavators. But this answer begs the question: Why was LTM able to employ mechanical devices? This is due to the fact that LTM had a large enough area, and one that was relatively proven. This tended to reduce the risk of losing one's capital since the recovery rate was relatively high. Thus the problems of the native firms hinged on a cyclical thesis. They did not have enough resources (capital and land) to engage in long-term prospecting which meant that they had to make do with small plots which in turn called for labor-intensive techniques which could not efficiently recover stones from the deeper deposits.

The Nature of Labor Under the ADMS

We have argued that the category wage-labor is not applicable to the mode of remuneration under the ADMS. If we compare the methods of recruitment and remuneration of tributors with the modern wage-laborers (within SLST), we find that one important respect in which they differed was the degree of economic freedom they enjoyed. Wage-workers freely entered into contract of employment with employers and were free to move from one employer to the other. By contrast, the tributor had to be coaxed out of the peasant family farm, thus suspending their usufructuary right to land until their return at the end of the mining season.[26] This meant that the tributor, unlike the wage-laborer, was not totally dependent on his employer in the diamond field for his livelihood. This fact affected the final benefit which the tributor received from his productive efforts, one which was only designed to provide for part of his reproduction.

Furthermore, percentage sharing of the final sale of the product of labor of the direct producer (with the tributor laying claim to these products), has nothing to do with capitalism. On this point Marx was quite unequivocal:

> ...the product is the property of the capitalist and not that of the laborer, its immediate producer. Suppose that a capitalist pays for a day's labor power at its value; then the right to use that labor power for a day belongs to him; just as much as the right to use any other commodity, such as a horse that he had hired for a day. To the purchaser of a commodity belongs its use, as the seller of labor power, by giving his labor, does no more, in reality than part with the use-value that he has sold. From the instant he steps into the workshop, the use-value of his labor power and therefore its use, which is labor, belongs to the capitalist. By the purchase of labor power, the capitalist incorporates labor, as a living ferment, with the lifeless

constituents of the product. From this point of view, the labor process is nothing more than the consumption of the commodity purchased, i.e., labor power; but this consumption cannot be effected except by supplying the labour power with the means of production. The labor process is between things that the capitalist has purchased, things that have become his property. The product of this process belongs, therefore, to him, just as much as does the wine which is the product of a process of fermentation completed in the cellar.[27]

If we look at the relations of production of all the "native firms" and individual mining ventures within the ADMS, we find one common feature, i.e., the method of remuneration. We have already noted that this has nothing to do with wage-labor, and hence the relations of production is not capitalist. On examination it resembles a version of the *metavage* system, in particular, the share-cropping system of the Southern United States.

The share-cropping system of the USA resulted from the need to link the soil and the cultivator. The former slaves lacked funds to buytland; the slave owners lacked funds to pay wages. Hence the agreement that the landlord would provide the land, the freedman the labor, and each would receive a share of the proceeds from the final harvest. The cultivator, like the tributor under the ADMS, received the necessities of life on credit during the annual periods between harvests. Similarly, we have argued that because alluvial mines were usually restricted to small plots in the marginally diamondiferous areas, it became very risky to try to employ tributors on a wage basis. We have seen that only in the most developed sector of the ADMS, namely *Leone Trial Mining*, because relatively small plots in proven grounds were consolidated into large leases, was it possible to employ skilled workers such as dragline operators in return for wages.

Diamond Mining and Underdevelopment in Sierra Leone

To suggest that diamond mining generated underdevelopment in Sierra Leone is not to deny that certain groups and individuals did not benefit from what was for some a lucrative operation. Many individuals, foreigners and Sierra Leoneans alike made windfall gains, the industry provided much-needed-foreign exchange earnings for the Government, it has provided wage and non-wage employment for thousands of Sierra Leoneans.

These gains are not being denied; instead three critical points are being emphasized. Firstly, that Sierra Leone or Sierra Leoneans did not do as well as other participants in the industry. Secondly, the damage done to the diamondiferous areas and to agriculture in particular is such that it will take years and a massive injection of capital to rehabilitate these areas. Finally, we try to show that the net effect of diamond mining was to prevent local people from taking the initiative to control their economy. We have also drawn attention to the fact that these negative features were largely the result of SLST's dominance in the industry.

The need to protect SLST's lease was the major reason behind the ordinance of 1936,[28] which empowered the Government to prohibit and restrict the residence of "strangers" and the issuing of store and other trading licenses in the diamond areas. Kaniki, who has done some study of this period in the country's history, has argued that it was through the application of this legislation that SLST's security force harassed non-Kono traders and others, and systematically reduced their numbers in the District.[29] The relative isolation of this area has led one writer to suggest that:

> The slow pace of economic change in the pre-war period is partly attributable to the deliberate policy of isolation practiced by the colonial authorities and the SLST after the discovery of diamonds in 1930.[30]

The prohibition of "strangers" from settling in the District was a clear interference in the traditional way of life of the people in the area. Furthermore, it blocked opportunities for local entrepreneurs who would have expanded the exchange sector.

The emergence of mining sector struck a serious blow at the agricultural sector, particularly in Kono where agriculture had to compete with mining for land. In practice the latter usually triumphed, mainly because of the long gestation period between planting crops like cocoa and coffee and profit realization, compared to the relative ease with which compensation could be obtained from would-be miners, and the fact that one good discovery in the diamond field is worth more than a lifetime of farming.

The increase in population resulting from migration into the diamond areas put further strain on the distribution of farmland. This led to overcrowding, with the result that many people had to leave the land in search of employment in the towns. Since jobs had always been hard to come by, this meant that many one-time farmers joined the ranks of the urban unemployed. It is true that some of these ex-farmers moved into the mining areas as tributors, and others got into the industry as gang-masters. Earlier, we have noted the fate which awaited these tributors which we have argued was the direct result of SLST's dominance in the industry. In this way, we can see how the production and profitability of SLST resulted in the marginalization of the mass of the people in Kono District. This process of marginalization has been made more acute by the mining policy of SLST. The company's operations have been carried out with the aid of heavy machines on upland and swamp areas in both Kono and Kenema Districts. Most of these mined areas were left to perish without any serious attempt at rehabilitating the land. This point has been emphasized by a field worker, who observed that:

> The conditions of the people in the Kono and Tongo leased areas is growing worse every day without proper and adequate rehabilitation...the mining by the company of the swamps and valleys has left the people in the leased areas to concentrate on upland rice and farming. This has caused a lot of strain in producing their staple food in any appreciable quantity...the people are now left with no alternative but to go to the hills, far from their settlement to farm. This (mining) has caused a lot of problems in these areas...lack of good farming land, lack of good drinking water, infestation of mosquitoes, rapid spread of diseases and low earning capacity of the majority of rural people in these areas.[31]

The need to protect SLST's lease from the activities of IDM posed further problems on the farming community in Kono. For example, permission was needed to cultivate plots close to or within the leased areas, and this was granted only if the areas were not marked for future mining. Fishing, a major source of protein, was prohibited in streams within the leased areas as diamonds could be found in river beds.

With all these constraints on farming matched by the demand for manual labor in the mining industry, it is not surprising that the percentage of farmers in Kono is the lowest of all the Districts in the Provinces. For example, in 1970, only 46% of the labor force was employed in agriculture as compared to between 70-75% for other districts.[32] This drop in the percentage of agricultural producers could indeed by seen as a sign of progress, provided that those who left the land were able to find paid employment in other sectors of the economy; and also provided that those who remained on the land were able to produce enough to feed the increasing number of non-producers. We have already noted that for most of those who left the land, paid employment was not a feasible proposition. With regards to the productivity of those who remained as farmers, Saylor has argued that as far as subsistence crops were concerned, there was probably an increase in productivity, since these did not suffer the same decline as export crops.[33]

Nonetheless, there were other factors which helped to negate this increase in productivity. These include poor communication, lack of infrastructure, and inflationary pressure - all factors which were more intensified in the mining areas.[34]

Who Benefited From Diamond Mining in Kono?

This question can be answered by looking at the benefits derived from diamond mining by major social groups within Kono. We can distinguish three such groups: 1) The Kono masses (i.e., urban/rural poor) who participated in "the pariah-like activities of illegal miners,"[35] 2) The rich and powerful (i.e., the chiefs and their political allies), and 3) SLST and other capitalist interests.

As far as the urban/rural poor are concerned diamond mining as we have seen tended to exacerbate their predicament. We have pointed to the paucity of employment possibilities outside of either tributing or illicit mining. We have also seen how the policy of protecting SLST's lease led to the persistent marginalization of the farming population in Kono. In terms of infrastructure, outside of the SLST enclave, Kono had the least infrastructural development of all the regions in the country.

In 1974-75, the Eastern Province which includes Kono and Tongo had the lowest number of secondary school pupils in the country: 8,333 compared to 18,684 for the Western Area; and 9,941 for the Southern Province. As far as hospital delivery system is concerned, in 1963 the Eastern Province accounted for only 14% of Government hospital beds, compared to 41% for the Western Area; 23% for the Southern Province and 22% for the Northern Province. Kono had the highest ratio of persons per medical officer (202,000:1) compared to 2,887:1 for the Western Area. This is also true of the ratio of persons per hospital bed, and also persons per health and medical center.[36] In short, Kono has served as an internal colony for the accumulation needs of peripheral capitalism in Sierra Leone.

If any indigenous group benefited from diamond mining it was the chiefs from the diamond mining areas and their political allies. This was possible mainly because the colonial authorities (and later the post-colonial rulers) and SLST needed their confidence in order to facilitate the exploitation of the country's mineral. The pretext for this approach was that since the chiefs were custodians of tribal land, they had to be recognized as such and that all decisions involving land should be channeled through these "natural rulers." Thus several agreements were signed between the chiefs in the diamondiferous chiefdoms and SLST, whereby the former agreed to exclude strangers (i.e., non-Konos) from their chiefdoms in return for stipends.[37]

By far the largest beneficiary of diamond mining was SLST and other capitalist interests in the industry. These benefits took the form of profit repatriation, smuggling and robbery.[38] The loss to the Sierra Leone economy due to profit repatriation cannot be estimated with any degree of certainty due to the paucity of trade statistics. However, by looking at one capitalist institution within the industry we can have a rough idea as to the gains made by metropolitan capitalist interests. From 1960-76, the Diamond Corporation West Africa limited (DCWA) managed the Government Diamond Office (GDO) and bought all the diamonds from the ADMS. During this period, a total of 13,881,039 1/4 carats were bought for Le419,971,622; and the Government of Sierra Leone received Le31,4 million in export duties and license fees. The DCWA received a total of Le54.6 million from its operations with the ADMS.[39] This figure does not take into consideration the profit made by the DCWA from its purchase and sale of diamonds in Sierra Leone, or profits made through deliberately buying stones below the scheduled price, i.e., by undervaluation.[40]

With regards to smuggling, I shall argue that the dominant position of SLST within the industry, which we have seen led to the supporter/tributor system, helped to encourage smuggling.[41] Furthermore, I shall try to show that the activities of the Diamond Corporation encouraged and facilitated smuggling.

Prior to the legalization of alluvial mining, two major smuggling networks had been established linking cutters in Europe and North America; through Beirut in Lebanon and Monrovia in Liberia.[42] The Beirut connection emerged as a result of citizens of that country illegally exporting Sierra Leone diamonds. In the case of Liberia, the crucial factor was the dollar premium which could be earned by exchanging diamonds in a country where the US dollar is a legal tender. According to Van der Laan, Liberia had no deposit worth mentioning; the success of the Liberian market was based on supplies mainly from Sierra Leone, Guinea, and Ivory Coast. Yet, this did not stop the Diamond Corporation from setting up an associated company in Liberia to run the Diamond Appraisal Office on behalf of the Liberian Government.

We now turn to look at two important questions: 1) Why did smuggling continue even after the establishment of the Diamond Corporation in Sierra Leone? and 2) Why did the DCWA want to enter the Liberian market?

Some of the answers to the first question include factors which were internal to the Liberian economy, such as the lower effective duty compared to those charged in Sierra Leone and the dollar premium. But a crucial reason was the trading policy of the DCWA. According to Van der Laan, in the period 1956-59, the trading margins of the company proved too high, with the result that buyers who were in a position to compare prices in the Monrovia market and at the DCWA soon found out that their profit margin would be reduced by making sales at the DCWA. This policy of very high trading margins was reversed in the period 1959-67. This change of policy raises the question why the Corporation wanted a stake in the Monrovia market. In 1959, when it took over the GDO in Freetown, it embarked upon a policy of buying as much as possible rather than to earn as much as possible.[43] This policy fitted in well with its strategy of maintaining as large a share as possible of the world's market of diamonds. However, this policy would succeed only if the Central Selling Organization (CSO) was prepared to forego more profit and accept a lower trading margin.[44] The CSO, which at the time had become very worried about the Monrovia market (over which it had no control), decided to make good its losses in the Monrovia market. This step was necessary because most of the good quality gemstones from Sierra Leone were channeled through this market, whilst only the poor quality industrial stones were sold through the GDO. The operation in Liberia simply meant that what was lost in Sierra Leone was gained in Liberia.

It is not easy to arrive at an estimate of the value of stones that left Sierra Leone for foreign markets. However, using Van der Laan's formula, in particular his assertion of 1:1 ratio of gem and industrial diamonds for Sierra Leone's deposit,[45] we can arrive at a rough estimate. This estimation is premised on the assumption that the industrial stones were not smuggled in any appreciable quantity, so that the volume of recorded industrial diamond exports were taken as representing 50% of actual production under the ADMS. We then subtract the volume of recorded gem exports from the volume of recorded industrial diamond exports to obtain an estimate of the volume of smuggled gems. Assuming the same unit per carat as for recorded gem diamond exports, the value of smuggled gem diamond exports was then obtained. The estimated value of smuggled gem diamonds averaged roughly Le15.5 million per annum over the 14-year period 1960-71.[46] The estimated value of smuggled diamonds represented 34% of the value of recorded diamond exports and 21% of the value of total domestic exports.[47] In effect the value of exports could have been boosted by at least one-fifth if effective measures were taken to combat smuggling. In addition, the loss of export duty averaged 33% of actual revenue from export duties over the period under review; and 2.5% of total Government revenue from all sources.[48] To the extent that the exchequer and the economy lost these revenues which could have been used in development efforts, smuggling tended to intensify the problem of capital drain and underdevelopment.

CONCLUSION

In this paper, I have tried to show how diamond mining led to the development of underdevelopment in Sierra Leone. In locating the modes of production within the industry I have argued that the structure of the non-capitalist sector was determined by the hegemonic position of the capitalist sector within the industry. Because most of the resources (capital, trained personnel, and rich deposits) that entered the industry found their way into the SLST sector, the company was able to exercise its hegemonic control over the industry. We also noted that this hegemonic position of SLST largely accounts for the atomized nature of the ADMS. The case of *Leone Trial Mining* showed that with relatively large and rich deposits as well as capital it was possible for relations of production in the industry to be transformed. Furthermore, we saw how the company tried to articulate its interests with those of pre-capitalist forces, such as chiefs.

With regards to the surplus that was retained within Sierra Leone, we noted that because of poor infrastructure in the mining areas and the absence of any welfare provisions, these areas could be said to have acted as an internal colony. We also argued that for the majority of the people in these areas, diamond mining has caused more harm than good. We drew attention to how agriculture had to compete unsuccessfully with mining for land. The net effect was that farmers had to cultivate upland areas with all the physical and social strain that this entailed.

NOTES

1. See A. Zack-Williams, "Merchant Capital and Underdevelopment in Sierra Leone," *African Review*. Vol. 10. No. 1. 1983.

2.. *Ibid.*

3. *Ibid.*

4. See the Alluvial Diamond Mining Ordinance and Rules, 1956.

5. *Ibid.*

6. As theorists of dualism have argued.

7. P. Raikes, "Rural differentiation and Class formation in Tanzania," *Journal of Peasant Studies*. Vol. 5. No. 3. April 1978.

8. H. L. Van der Laan, *Sierra Leone Diamonds*. Oxford, 1965, p. 73.

9. *Ibid.*, p. 35.

10. For details of the conflict, see Van der Laan, *ibid.*

11. See my "Merchant capital and underdevelopment in Sierra Leone," *op. cit.*

12. Van der Laan, *op. cit.*, p. 78.

13. *Ibid.*, p. 27.

14. D. B. Rosen, *Diamonds, Diggers and Chiefs: The Politics of Fragmentation in a West African Society,* unpublished Ph.D. Thesis, Urbana-Champaign, 1973.

15. See my *Underdevelopment and the Diamond Industry in Sierra Leone,* unpublished Ph.D. Thesis, Sheffield University, 1980.

16. This was set up in 1960 with a grant of $100,000 from the American Government to assist miners to purchase mining equipment.

17. Zack-Williams, 1980, *op. cit.*, p. 205-207.

18. M. S. Deen, *An Appraisal of the ADMS,* Mines Division, Kenema, 1972.

19. T. L. Balewa, "The system of reward of labour under the ADMS," Government Engineer's file, Mines Division, n.d.

20.   But he did not own the land.

21.   Since they owed their protection to the gang-master.

22.   The social relations between the tributor and the gang-master is one of client-patron relationship.

23.   Balewa, *op. cit.*

24.   The size of holdings vary according to the type of mining to be carried out; river license restricted to 2000 ft. x 50 ft. and land license to 200 ft x 200 ft.

25.   For a detailed discussion of the project see U.B. Usman, "Leone Trial Mining: Mimeo Mines Division" Koidu, n.d.

26.   The mining season coincided with the dry season when farming activities are least demanding.

27.   K. Marx *Capital*, Vol. 1, Lawrence and Wishart, London, 1974, p. 180.

28.   Chapter 196 of the *Laws of Sierra Leone 1960*, Section 3, 1.

29.   M. H. Y. Kaniki, The Economic History of Sierra Leone; 1929-39, unpublished Ph.D. Thesis, University of Birmingham, 1972, p. 295.

30.   V. Minikin, *Local Politics in Kono District, Sierra Leone, 1945-70*, unpublished Ph.D., University of Birmingham, 1971, p. 145.

31.   S. Dumbar, "Rehabilitation in the NDMC (S.L.) Ltd, Leased Areas in Kono and Tongo Field," Mimmeo, n.d., Mines Division Kono.

32.   Figures from the *National Accounts*, Central Statistics Office, Freetown, 1975.

33.   Saylor, *op. cit.*

34.   "Household Survey," *National Accounts 1970/71-1975/76*, CSO, Freetown.

35.   Minikin, *op. cit.*

36.   Figures from *Directory of Medical Units*, Ministry of Health/WHO Team, Ministry of Health, Freetown.

37.   Minikin, *op. cit.*

38.   M. Harbottle, *The Knaves and Diamonds*. London, Seeley, 1976.

39.   Figures from Mines Division, Kenema; and Zack-Williams *Underdevelopment and The Diamond Industry*, *op. cit.*, pp. 280-282. Sierra

Leone Government share is equivalent to 7 1/2% duty imposed on sales at the GDD; DCWA's consists of 12% which SLST paid to the Central Selling Organization and the 1% which it charged as commission for managing the GDD on behalf of the Government.

40.   See Zack-Williams, 1980.

41.   Smuggling in this context refers to the patronizing of overseas markets at the expense of the locally instituted market.

42.   Van der Laan, *op. cit.*

43.   Van der Laan has noted that since the discovery of the Sierra Leone fields, De Beers (the parent company of DCWA) had been anxious that this field might upset its control of the world diamond market. These policies must be seen in that light.

44.   The CSO has been the major cartel in the industry since the 1930s and is largely controlled by De Beers.

45.   Van der Laan, *op. cit.*, p. 134.

46.   This estimate was arrived at from figures from the bank of Sierra Leone, *Economic Review*, July-December, 1974.

47.   Zack-Williams, p. 305.

48.   *Ibid.*, p. 305.

CHAPTER VII APPENDIX I

Table Comparing Production of S.L.S.T. and Contract Mining 1965-1970

Year	S.L.S.T.	1965	1966	1967	1968	1969	1970
1.	Overburden moved (Cu yds)	3,346,528	3,399,751	3,822,360	4,093,586	5,791,526	7,923,292
2.	Gravel treated (Cu yds)	759,714	728,477	812,273	866,883	1,200,743	1,444,996
3.	Carats recovered (in Carats)	652,410.1	701,894.5	660,455	655,723	831,488.4	988,196.7
4.	Grade of Gravel (Carats per cu yd)	0.86	0.96	0.81	0.76	0.69	0.69
5.	Average Overburden (Depth ft.)	8.6	9.0	8.8	10.5	10.6	12.40
6.	Average Gravel (Depth ft.)	2.2	1.9	2.0	2.3	2.1	2.2

CHAPTER VII APPENDIX II

Contract Mining

Year		1965	1966	1967	1968	1969	1970
1.	No. of sites	11	16	11	3	2	2
2.	Overburden stripped (cu. yds.)	150,649	291,566	57,715	27,521	7,069	-*
3.	Gravel treated (cu. yds.)	33,466	61,260	15,118	6,788	537	-*
4.	Production (Carats)	19,660.20	23,638	8,630.70	1,484.60	N.A.O	-*
5.	Grade Ct/Cu.yds)	0.77	0.39	0.59	0.22	N.A.O	-*
6.	Value per carat (Le)	13.22	9.32	11.45S	16.20S	N.A.O	-*
7.	Cost per Carat (Le)	8.08	6.98	4.06S	2.20	N.A.O	-*
8.	Employment	682	650	730	160	328	-*

*-Contract Mining was stopped in March 1970
O-Figures not available
S-Applicable to end of June
Source: *Report of the Mines Division*, Ministry of Lands and Mines, Freetown.

Chapter VIII

Glass-making Technology in Nupeland, Central Nigeria: Some Questions

Gloria Thomas-Emeagwali
and
Aliyu Alhaji Idrees

Introduction

In this chapter we first of all examine some of the qualities and identification criteria of glass and various methods used over time in its fabrication. In the course of discussion we reflect on some of the features of Egyptian glass-making techniques since this would enable us to evaluate to some extent the claim of Egyptian origin on the part of Nupe glass producers. We then identify aspects of glass-making technology in Nupeland particularly in terms of centers of production, changes in techniques over time and raw material availability. We conclude by reflecting on some of the several unanswered questions on glass-making technology in the area of focus identifying areas of priority for archaeological research.

Identification Criteria of Glass

Glass is an amorphous solid, often regarded as a super-cooled liquid rather than a true solid.[1] Several features are associated with it, some of which are rigidity, high viscosity, and the fact of its being a non-crystalline substance. The term glass has its origin in the Germanic term but as pointed out elsewhere this in itself is by no means an indication of prior invention by

Germanic peoples.[2] The oldest term for glass is Egyptian and as we shall discuss further in the course of this paper, glass-making technology in fact diffused from Egypt to Europe through a process of technology transfer.[3] One of the most important principles governing glass-making is that molten glass must be slowly cooled since the process of annealing is vital. When cooled slowly, the fused material becomes more viscous, congealing to a hard solid. If it is not cooled slowly, de-vitrification takes place: The silicates crystallize. The glass loses its transparence and easily falls to pieces.[4] It is also important that there should be no impurities in the raw materials used or the glass produced would not be transparent. If insufficiently heated, the glass produced manifests bubbles, indicating that there are trapped pockets of air. Glass may also undergo a process of internal decomposition over time if the process of annealing were improperly done.[5] Iridescence and the display of variegated coloration and rainbowlike colors constitute another feature of glass defective in manufacture. The making of glass entails the fusion of alkalis or their salts, lime and silica (sand). The proportion of each component is important. If there is too much silica or lime, the glass produced is opaque. Too little lime produces a glass that may become porous. If the percentage of alkali is high it makes the glass more corrosive. In the case of glass-making in Nupeland, it is important to find out the extent to which glass-makers were able historically to guard against some of these problems. In the discussion which follows, we focus on the Egyptian factor with respect to the global development of glass-making technology.

Egypt and Glass-making Technology

The casting of glass into separate objects first started in Egypt at least 3,500 years ago, making Egypt one of the earliest glass-producing centers. It became a principal source of technology transfer to Europe. Even so, as late as 675 A.D., there were formal requests from England for glass-makers since it was admitted that 'the art was unknown in this country.'[6] Forbes has also alluded to the role of Syrian immigrants in the process of transfer. Many Syrian as well as Alexandrian glass factories were set up in Italy. The process of technology transfer from Egypt (and also Syria) to Greece was less successful. The Greeks failed to develop a significant glass industry either by indigenous development or technology transfer. Technology diffusion from Egypt to Europe was facilitated by the Roman occupation of Alexandria, Egyptian migrants to Rome and the receptivity of the Gauls and Franks to Egyptian and Syrian methods but, even so, as pointed out by Forbes, not until the 4th century did Gallic glassmakers get away entirely from Alexandrian influences in techniques and shapes. Venetian importation of vitreous earth and natron from Egypt persisted long after the period.[7] Glass-making technology is therefore a classic case of the process of technological diffusion from South to North in a foundation period of science and technology.[8] We have spent some time on this issue because as we shall see, Bida glass producers also claim Egyptian origin in their techniques. It may be useful to conclude this section therefore with a brief focus on the principles of coloration and the chemical composition associated with the Egyptian products since this would facilitate our appraisal of the Bida claim, or at least provide some preliminary suggestions for future research into this issue. Generally speaking the color of a glass product is determined by the metallic oxides associated with its manufacture. Copper, cobalt, lead, manganese copper and lead antimony compounds, for example, are associated with blue, violet, yellow, black and yellow, respectively.[9] In the Egyptian case blues predominated. Egyptian glass was essentially a soda-lime glass rather than a potash-lime glass as pointed out by Forbes and was therefore easier to mould than the latter. In the case of the soda-lime glass, natron or sodium carbonate is used whilst potash-lime glass consists of ashes obtained from wood. Needham points out that Ancient Chinese glass consisted of a high percentage of lead and barium. This gave it a high refractive index and a lower melting point than the soda-lime or the potash-lime glass.[10]

Glass-making Techniques

Historically, glass has been made by one of several methods. The fused and molten mixture of salts, lime and sand is poured into a mould and pulled out as threads or, alternatively, the mixture is allowed to cool and broken off from the container or the molten liquid blown into shape by a blow-pipe. Another method, the fourth to be mentioned here involves the remelting of glass and the reworking into glass of the molten mixture so produced.[11] We may point out here that in some areas, glass production was preceded by the production of glaze or faience, which do not really vary in terms of chemical composition. In the case of glaze the molten material is used to cover other objects whilst faience is basically the application of the molten material to pottery. It would be useful to know the extent to which Bida glass-producers are familiar with the making of glaze and faience and whether there is any evidence that they made these before they produced glass. We shall reflect on these questions in the discussion which follows, but before this we comment briefly on the significance of the furnace in glass-making technology.

We may distinguish between technologies which involve high temperatures and those which do not.[12] Glass-making technology ranks amongst those such as metallurgy and ceramics, which do indeed require high temperatures and which therefore need efficient furnaces. There are those which need a combustion chamber from which hot gases passed into another chamber containing the matter to be heated, and others in which the material to be heated mixed with the fuel.[13] Glass-making technology often employed the former type of furnace. Taylor and Singer have pointed out that the attainment of high temperature depended on the type of fuel used and the draught obtained, and that the latter was largely related to the effectiveness of the bellows used and the actual siting of the furnace.[14] In our focus on glass-making technology in Bida we shall reflect further on some of these issues. We may note, however, that in the case of the Egyptian furnaces, especially the early ones, a hearth pit was dug. There were also cases of beehive-shaped hearth furnaces with one or two compartments for a small crucible and one or two small compartments left empty to take the blown glass objects for cooling.

Glass-making Technology in Nupeland

Location of Glass-works

The city of Bida is located about 50 kilometers from the Niger-Benue confluence and constitutes one of the prominent historic centers of the Nupe-speaking peoples. It is at present one of the densely populated cities of the

Savannah region of Nigeria.[15] Since the middle of the 19th century, Bida has been a center for various industries. As pointed out elsewhere it was a center of textile production both in terms of cotton and silk fibers.[16] Colonial authorities identified 1,656 tailors in the city and over 1,600 inhabitants of Bida were employed in the making of hats and mats in 1921.[17] The Colonial Agricultural officer, Bida, in a letter to the Resident, Niger Provincial Office, Minna, declared as follows:

> Most of the pre-war established local industries such as glass, bangles, mats, pots, brassware, hats and cloth have always been in good condition without any aid from us...the least we interfere with this natural development the better....[18]

At present glass-producers are quartered at Masagafu within Bida city, where they have their workshops.[19] According to oral tradition, the first place or settlement of the glass-workers however was Gbara, an administrative capital in pre-19th-century Nupeland. Nupeko, Zhima, Lade and Raba have variously served as capitals in the Nupe-speaking region and glass-producers and their workshops are said to have been associated with these centers at various periods.[20] Raba, the last capital before the Sokoto Jihadists established Bida as their administrative base in Nupeland, served as one of their centers in the pre-19th-century period. The shift of glass-producers to Masagafu took place after 1857, the period of the foundation of Bida as an emirate headquarters in the context of the Sokoto Caliphate.[21] Oral evidence in fact suggests that Emir Masaba of Bida, during his reign from 1859 to 1873, personally encouraged and supervised the resettlement of the glass-workers at their present abode.[22] We may note, however, that one tradition suggests that immediately before this time the glass-producers were located at Jemaa. Another suggestion identifies Raba in this regard.[23]

Guild Organization

Several guilds were established in Nupeland, each with their chief administrator and leader, and designated specific titles and modes of salutation. There were guilds of tailors, carpenters, brass-workers, iron-workers, dyers, leather-workers, bead-polishers and also glass-makers.[24] In the case of the glass-makers, the head of the guild was greeted as Turawa, a salutation which was designated to glass-makers in general and therefore all members of the guild. We should note that in the perception of indigenes of the area the term Turawa refers to North Africans. The extent of power and control wielded by senior officials of the guild over other producers is an aspect of labor history that needs further research. It is clear, however, that it was difficult for a member to leave the guild and set up a workshop independent of its organizational control. It may be noted as well that apprenticeship and training in glass technology was restricted only to descendants of earlier glass-producers. In this sense there was the emergence of a caste of glass-producers within the generality of Nupe manufacturers and technologists.[25]

The role of religion in ensuring loyalty to the guild can hardly be overestimated. Charms were buried in the ground by the elders before a new glass factory was constructed and it was a common belief that no furnace could be successful in the absence of this activity and their consent. We may note here that in the Egyptian case the high priest of Memphis bore the title of 'the director of the glass factory.'[26] If the glass-producers were in fact migrants from Egypt as is generally claimed by the glass-producers, then the use of religion by Nupe glass-producers is simply a continuation of an earlier practice. We shall reflect more on this issue of "origins" in the concluding section of this work.

Glass Technology at Bida

As mentioned earlier glassworks have been established in various parts of Nupeland over time. Their establishment at one place or another was influenced by political and administrative developments although it is not clear as to whether the glass-producers, in particular the upper echelons of the guild hierarchy, were themselves directly involved in politics at the level of the state apparatus. Our knowledge of glass-working technology in Nupeland is largely derived from oral tradition and investigations emanating from Bida, the most recent area of settlement of glass-workers, dating back to 1857, as mentioned earlier. Two types of techniques have been identified in this area. The first type has to do with the old and traditional method of glass-production in the region. The second type is a blend of the old technique and method with more recent innovations, stemming from changes in the raw material used. We may associate this with the fourth technique of glass-making earlier discussed. It may be observed that in the case of the first type there is the exclusive use of locally procured raw materials which include sand, potash or natron and charcoal. It would seem that the area produced two of the types of glass referred to earlier, namely, a soda-lime glass as well as a potash-lime glass. It would seem that in the pre-20th-century period there was the predominance of the potash-lime glass.

Glass producers claim that sand collected from any part of Nupeland was appropriate for glass manufacture, a conclusion contrary to that of Jefford. The latter, in his survey of sands near the Mokwa-Bida ferry found cases of irregular grain size and pointed out that they were not ideal for smelting particularly because large grains took a long time to react whilst very fine ones entrapped air bubbles.[27] Screen analyses of sand samples from river terraces within a two-mile radius of Bida yielded a fairly similar report.[28] Chemical analyses of the samples indicated their suitability for the production of colored glass rather than colorless ware, given the iron content.[29] We may note excerpts of the report:

> *Bida, Niger Province.* Sand samples collected from river terraces near Bida proved to be badly graded with much coarse material. The 'ideal fraction' ranged from 42 to 58 percent. The grading of two of the samples could be improved by a combination of crushing and screening but the grading of a third would lead to the introduction of appreciable iron from the coarse laterite fragments. All 36-100 mesh fractions are suitable for the manufacture of colored bottle glass and one sample with an iron oxide content of 0.14 percent meets the requirements for ordinary window glass....[30]

From the above it seems that glass-producers in Bida have been relatively disadvantaged in terms of accessibility to ideal sands. Their accomplishments so far in glass manufacture must be fully complimented and acknowledged.

We have alluded to one of the methods utilized by glass-producers in the manufacture of glass and pointed out that the glass-makers produced both a soda-lime and a potash-lime glass. We may note that natron, a raw material in the soda-lime glass, was available from itinerant Hausa traders from Kano and Maiduguri. The same applied to potash. Having obtained the raw material, that is the appropriate sand, and the potash or natron, as well as firewood for smelting, an appropriate quantity of water was added and the mixture taken to the workshop furnace. The furnace itself was dug deep into the ground and was essentially a hearth-pit furnace, similar to some of the Egyptian furnaces. After the mixture was poured in, a high fire was set. The process of heating continued for two days until a vitreous liquid was produced. The mixture was then removed from below the furnace and allowed to slowly cool and solidify. The product was identified locally as 'bikini.'[31] The first stage in the production of glass was completed. It was now the turn of those who specialized in the transformation of the semi-processed material into the finished product. Quantities of the semi-processed material were therefore taken to other workshops where appropriate furnaces were used. There was thus a marked division of labour in the production process associated with this method of glass manufacture. We have elsewhere commented on the historian's dilemma in the identification of technical changes taking place over time in terms of local technologies. Glass technology in Nigeria invites a similar comment.[32]

The second method utilized by the glass-producers of Nupeland constitutes a deviation from the above. This method involves the use of bottles and glass objects, which are re-smelted and refashioned into the required consumer product. It is a development associated with the colonial period and a consequence of the influx of relatively cheap glass products from Europe.[33] Indeed it was estimated by colonial authorities in 1957 that whilst it cost 5d to produce a bottle in Nigeria, it cost 2d in the U.K. for a similar product.[34] In the context of an unprotected market, it was therefore more feasible for local glass-producers to utilize this method if they were to compete somewhat favorably with the fully imported product. We may note that a similar development took place in terms of metallurgy with the introduction of the 'scrap policy', a policy which features prominently in the decline of traditional metallurgy in Nigeria. In this case, alternative metal designated as 'scrap' was imported into the region from Britain. Abubakar has argued that this policy was a deliberate attempt on the part of the British whereby local products were to be replaced by those from European factories and that the policy succeeded in strangulating the local metallurgical industry and arrested the further development of indigenous technology, particularly in terms of the Zamfara metal industry, Sokoto.[35] This situation is hardly different in terms of glass-making technology. We may note that colonial authorities imported beads worth $107,640 in 1950 and $486,985 in 1952,[36] a substantial amount for that period. We have not come across the estimates for bottle importation. In the closing years of colonial rule there were attempts to investigate Bida glass manufacture and its potential in recognition that in the region there was "a small native industry of some long standing."[37] Glew admitted that one of the major problems was the competition of 'cheap imported machine-made glassware.'[38] These attempts, however, were much too late and at any rate were undermined by various other developments, some of which included the activities of N. Azikiwe in his quest to establish a glass-making factory in Enugu, Eastern Nigeria. By 1959, the eve of flag independence, three experts had actually visited the latter region to do feasibility studies on the issue. We note the following comment:

> ...Easterners have found sand white enough to make clear glass and (that) a factory is probably on with the American technical partner, it will prevent any Northern enterprise.[39]

Unanswered Questions

The development of glass technology in Nupeland is one of the various under-researched issues in Nigerian history.[40] There are several unanswered questions in need of archaeological research. As pointed out earlier, the earliest glass-making sites in the region may be located in Gbara, the area claimed to be the first port of call in the region of migrant glass-makers. It would be useful for us to know the design, size and capacity of the early furnaces; the raw material identified in glass artifacts; the extent of decomposition, and iridescence of the glass made at these sites; and most important the date of production. If there are remnants of glaze and faience we may postulate that the techniques of the glass-producers evolved no differently from those of other regions and this may influence our perception of the origin of glass-making technology in the area. We would benefit from knowing whether there are changes in raw material, techniques and overall style in the glass produced at other sites such as Nupeko, Zhima, Lade and Raba, other areas of settlement of glass-workers in earlier times. Laboratory analysis of specimens from the diverse sites would inform us as to whether the coexistence of potash-lime and soda-lime glassworks is associated only with production in this present century, or whether the trends predate the emergence of Bida as a glass-making center. A more careful analysis of the various types of instruments of production over time would also be facilitated and we should be able to match these with various Egyptian designs over time. Knowledge of the precise period from which there is the division of labor in terms of the production process would become more feasible, since we would try to find out whether all aspects of glass-making were done in a single furnace or whether auxiliary furnaces were developed to deal with the final stage in glass-making, as is the case in contemporary Bida.

100,000 beads, some beaded armlets and bead-studded head-dresses were amongst the artifacts found in the Igbo-Ukwu excavation. The beads were of various colors ranging from dark green to reddish brown. Venice, India and the so-called near East were the areas suggested as possible sources of the beads.[41] Ironically, no one thought of Nupeland as a possible source. But what if archaeological excavation in the region unearthed objects dating to the tenth century, or a period before this, in any of the early sites of the glass makers? Who were the earliest glass-producers in the Nupe-speaking region? Were they of Egyptian or Middle-Eastern origin as is sometimes claimed, or were they in fact full-blooded sons of the Nigerian soil? Masaga Bida emphasized that their forefathers migrated to Nupeland from Egypt, an observation endorsed by other Nupe-speaking groups within the city. Alhaji Masaga claimed that the migrants came to Nigeria through Borno, Kano and Zaria and then to Gbara, Nupeland.[42] We may point out here that there was constant migration of glass-producers from Egypt to regions as far as Europe, as earlier discussed. If a process of technological transfer took place from Egypt (and Syria) to Europe, in terms of glass-making technology, then why is a similar process of technological diffusion improbable in the case of Nigeria, a relative stone's throw away? If the first group of glass-makers in Nupeland were in fact of Egyptian origin, when in fact did they leave Egypt? Are they migrants from Egyptian Antiquity or are they products of an Islamized Egypt of the Medieval era? If, however, research suggests that the early glass-producers in the Nupe-speaking region were in fact wholly indigenous to the Nigerian region, then Nupeland must rank amongst the few areas of the world which manifested a largely indigenous effort with respect to glass-making technology. The sooner archaeologists assist in solving these historical puzzles the better.

NOTES

1.   A. Goodman and E. Payne, *Longman Dictionary of Scientific Usage*, Longman, Essex, 1981, p. 140.

2.   See G. Thomas-Emeagwali, "Technology transfer: Early Europe and the Third World," *Science and Public Policy*, International Policy Foundation, London, June, 1989.

3.   R. Forbes, *Studies in Ancient Technology*, Vol. v, Leiden, E. J. Brill, 1966.

4.   McGraw Hill, *Encyclopedia of Science and Technology*, 1977, N.Y., Vol. 6. See also D. B. Harden, "Glass and Glazes," in C. Singer et. al., *A History of Technology*, O.U.P., Vol. ii, 1972.

5.   Forbes, *op. cit.*

6.   *Ibid.*, p. 116, 164.

7.   *Ibid.*, 160.

8.   See G. Thomas-Emeagwali, "History and the question of technological development: The transfer of Technology revisited" in *Afrique et Developpement*, CODESRIA, Senegal, Vol. xii.2, 1987.

9.   Harden, *op. cit.*

10.   J. Needham, *Science and Civilization*, in China, Vol. ix, C.U.P., Vol. iv. 1972, reprint.

11.   Forbes, *op. cit.*

12.   S. Taylor and C. Singer, "Pre-scientific industrial chemistry," in Forbes, *op. cit.*

13.   *Ibid.*

14.   *Ibid.*

15.   *Cities of the Savannah*, Nigeria Magazine, Lagos, n.d.

16.   See G. Thomas-Emeagwali, "Reflections on the Evolution of Science and Technology in Nigeria," *Tarikh*, Vol. 1, 1989.

17.   National Archives, Kaduna, NAK Minprof 179/1909, Historical notes on Niger Province.

18.   NAK Minprof, 3183, Local Industries.

19.   Masagafu is the Nupe term for the quarters where glass-workers dwell exclusively. The term is derived from Masaga meaning glass worker.

20.   Nadel, *A Black Byzantium,* London, 1942.

21.   *Cities of the Savannah, op. cit.*

22.   Alhaji Saidu Madawaki Bida pointed this out in an interview held July 1988. March, 1989, reiterated this point.

23.   NAK Minprof 279/2909, Historical Notes on Niger Province.

24.   *Ibid.*

25.   *Ibid.*

26.   Forbes, *op. cit.*

27.   See NAK/Kadmin J.J. 73A   Manufacture of Glass and Glassware: General.

28.   *Ibid.* This was not really a problem for the local producers in the light of local tastes and the types of consumer products fashioned. See also Muyibi et. al., "Ferruginous groundwater treatment with bida sands," paper presented at the 15th WEDC Conference, Water Engineering and Development in Africa, Kano, March 1989.

29.   See Appendix.

30.   NAK/Kamin J. J. 73A.

31.   Oral interview with Alhaji Masaga Bida between the 10th and 20th March 1989 at Bida, Niger State, Nigeria.

32.   See Gloria Thomas-Emeagwali, "Oral Historiography and the History of Technology in Nigeria," in *MUNTU*, Revue scientifique et culturelly du CICIBA Centre International for Bantu Civilization, Presence Africaine, Paris, 8.1, 1988.

33.   NAK Minprof 279/1909.

34.   NAK Kadmin J. J. 73A.

35.   See N. Abubakar, "Colonialism and indigenous technology: The case of metallurgy in Northern Nigeria," in G. Thomas-Emeagwali, *Studies in the History of Science and Technology in Nigeria* (forthcoming).

36.   *Ibid.*

142

37.   NAK Kadmin J. J. 73A.

38.   *Ibid.*

39.   *Ibid.*

40.   See also other areas as discussed in G. Thomas-Emeagwali, "Alternative perspectives on the reconstruction of the African past," *Kiabara*, Vol. 4.2.   1981;   Thomas-Emeagwali, "Political institutions in pre-19th century Nigeria," *ODU*, no. 26. 1984; Thomas-Emeagwali, "The Woman Question in pre-capitalist socio-economic formations: The case of Northern Igboland, Eastern Nigeria at the end of the 19th century," *Ikenga*, Vol. 8.1, 1986.

For a discussion on areas in need of archaeological research, see Thomas-Emeagwali, "Notes on the History of Abuja, Central Nigeria," in *African Study Monographs*, Kyoto University, Japan, 9 (4): 191-196, February 1989.

41.   See T. Shaw, *Igbo-Ukwu*.

42.   Oral interview with Alhaji Yahaya Masaga Bida, at Bida, Niger State, Nigeria between the 10th and 20th March, 1989.

APPENDIX

Table I

B.S. Sieve No.	Size Range Inches			RT. 98	RT. 99	RT. 100	RT. 101
	+ 16		+ 0.0395	13	6	4	12
- 16	+ 25	-0.0395	+ 0.0236	19	12	10	17
- 25	+ 36	-0.0236	+ 0.0166	19	15	16	17
- 36	+ 100	-0.0166	+ 0.0060	42	48	58	42
-100	+ 120	-0.0060	+ 0.0049	4	7	6	5
-120			-0.0049	3	12	6	7
Total				100	100	100	100

Screen Analyses of four 20-pound samples of sand collected from river terraces within a two mile radius of Bida were examined as to their suitability for glass manufacture. British Standard Sieves were shaken for a period of twenty minutes on a Rotap Screen Shaker.

Table II

Screen Analyses Recast to the Specifications of the Society for Glass Technology

Fraction	Soc. Glass Tech. Spec.	RT. 98	RT. 99	RT. 100	RT. 101
Remaining on 16-mesh sieve	per cent Nil	13	6	4	12
Remaining on 25-mesh sieve	Not more than 1	32	18	14	29
Remaining on 36-mesh sieve	Not more than 5	51	33	30	46
Passing 100-mesh sieve	Not more than 10	7	19	12	12
Passing 120-mesh sieve	Not more than 5	3	12	6	7

Table III

Results of Chemical Analysis

Sample No.	SiO_2	R_2O_3	TiO_2	Fe_2O_3
RT. 98	94.00	1.68	0.29	0.31
RT. 99	97.94	1.53	0.14	0.19
RT. 100	98.16	1.56	0.14	0.14
RT. 101	98.02	1.53	0.13	0.19

Analysis for SiO_2, R_2O_3, TiO_2 and Fe_2O_3 have been made on the ideal fraction (-36 + 100 mesh) from each of the samples.

Compiled by Gloria Thomas-Emeagwali from NAK J. J. 73-A.

Chapter IX

Science and Technology Policies in Zambia

Donald Chanda

One cardinal feature in an industrialization program is the application of science and technology to production. The advent of the industrial revolution was in fact a product of such a combination. Colonial miseducation and colonial industrial policies left Zambia with inadequate scientific and technological manpower and physical facilities in terms of 20th-century science and technology. Whatever the colonialists had set up was to serve their interests in commercial farming as can be seen from Appendix 2. The National Council for Scientific Research (NCSR) was formed in 1967, in the post-colonial era, and was charged with planning a science and technology policy to carry out and coordinate basic and applied research in the country. It became operational in 1970.

In spite of its young age, the NCSR has scored a number of notable achievements. These include physical expansion of facilities and the training of a relatively large number of scientists and technologists. It has trained an equally impressive number of industrial processes and products, developed and released to industry and to pilot plants (see example in Appendix 3). However, its work, like the work of other R&D institutions in this country, has been inhibited in a number of ways, the most important being lack of funds from both Government and industry, and this has marginalized the particular role of R&D work in the development program. On the part of the Government policy-makers, it may be that the idea of spending money on experiments with a long gestation period (which R&D has) has not been recognized as being important enough.

Instead of giving it a central role and a full-fledged ministry, the Government views R&D activities as an appendix to its expenditures. A combined Ministry of Higher Education, Science and Technology dates back to 1988, and it must be noted that this notion, plus the commonly held but also mistaken view that money spent on unsuccessful research projects is money wasted, needs to be removed. On the part of industry, the wish for quick money encourages ventures in direct production and not so much research. Given the New Economic Recovery Program (NERP) this idea must fail because of the need for development and utilization of local raw materials. As for industry, the time has come when some importations are no more. In fact more and more industries are turning to scientific research to look into their possible development frontiers using local resources. The lack of a dynamic link between R&D and local industry will be solved given the aforesaid situation. This will also awaken R&D institutions to sound levels of management as they become more and more open to national programs. The tasks required in NERP also call for a mechanism for calling upon all scientists and technologists in R&D to make a consolidated effort.

The translation of scientific and technological information into operating technical systems is usually done by consulting engineering firms. The quality of products is also inspected by the National Standards body. Both are critical in the implementation of industrial infrastructure. These key instruments of a science and technology policy are a missing factor in Zambia. Many consulting engineering firms are subsidiaries of international firms. This vital capability is needed especially for the adaptation of imported technology. The present engineering workshops and foundries in the country can act as the nucleus for creating such a capacity, especially that of Tanzania-Zambia Railways - TAZARA, Zambia Railways and the Mines.[1] The Zambia Bureau of Standards (ZBS) has been enacted and

1. *Report on Research trip to Engineering Workshops,* September, 1986.

empowered to check against abuse of standard and quality of products. It cannot do much at the moment because if its limited physical infrastructure. For example, it cannot check the quality of brake shoes on the trains because it has no steel-and iron-or metal-testing facilities.

NERP and Science and Technology Policy

Among the well-intentioned demands of Zambia's NERP is the clause or requirement for the manufacturing industry to depend on local raw materials. The utilization of existing industrial capacities, promotion of inter-linkages between agriculture and manufacturing, and selection of technologies are among the central objectives in the NERP.[2]

In most fields the initial effort will be towards the adaptation of the technological ware to local conditions, i.e., manpower, machines and materials. While acquiring selected technology from abroad, this country needs to develop indigenous research capacity. Indeed without such local skills, the process of adapting the technology would be inordinately expensive and time consuming and would remain incomplete. While concentrating limited resources on applied technological R&D, we need to focus on development priorities for providing food, housing and health services to large rural communities.

The R&D institutions must establish priorities in line with the industrial strategy to ensure maximum efficiency. The priority today is utilization of local raw materials in industry, hence it is just in order to explore the availability of local raw materials through an inventory (Resource Inventory), which should be updated periodically. Their utilization in industry through necessary processes, whether by modification of machinery or diversification of production lines, is but the next logical step. A reorganization exercise will follow this where researchers become accountable to national projects and objectives, not to peer groups, while the

---

2.   *Republic of Zambia: New Economic Recovery Program Interim. National Development Plan,* Government Printers, Lusaka, 1987.

government, parastatal and private firms refer their industrial problems and consultancy to the scientific institutions.

The confidence between the two has had to be created and consolidated. The government, like industry, should have a clearly articulated perception for R&D and should give it due priority in terms of finances and material equipment necessary for research. Moreover, a survey of manufacturing industries conducted by NCSR to identify the research and technical needs revealed that companies need R&D support.[3] As incentives, holidays or new investments in priority areas, especially using locally developed materials and technological processes, must be encouraged. This will go a long way in furthering the use of local R&D results in industry and cement the much-needed productive inter-relationship. The Ministry of Higher Education, Science and Technology recently set up should not mean tying up R&D activities in bureaucratic red tape but rather should act as a lubricant to make R&D work smoothly without much friction with industry and give R&D its much-needed national primary. Among its activities could be to see to it that sectorial companies with common research needs finance their research either through NCSR or in their own research units but with knowledge and consultancy of NCSR research units, even agriculture research units lump-sum grants. They in turn should be free to consult the R&D units whenever they have a particular problem. Let the R&D units at NCSR, Ministry of Agriculture parastatal and private companies operate independently. This will foster competition within the R&D units for industrial recognition where it does not exist and make R&D grow with industry. Such competition within cooperation is not harmful to national development.

A new, more-explicit patents and industrial-licensing system must be followed if we are to break off our technological dependence.

Let me use the example of S. Korea to support my argument. Korea has managed to reach the level of technological development that it has today because of its ability to modify and adapt the technology. The country has developed its own way of using technology and capital goods through

3. Peter Mwamfuli, *Research Report on Manufacturing industry* R&D *and Technical Services Needs* NCRS, Lusaka, 1985.

what is known as indigenous production engineering. Machiners and technological application have been handled in simpler ways than originally designed since workers are relatively poorly equipped with skills. As long as the simpler way of doing a thing has resulted in the desired product with higher values added, not only have output goals been reached but Korean management and workers have gained confidence and have also undertaken further improvement and adaptation of imported technology. In this way Korean industries have expanded their supplies of capital while utilizing abundant labor.

The much-needed prerequisite for all these policies to be implemented is an explicit national science and technology policy in which long-term strategic goals are set and medium-term opportunity areas identified. Special inter-ministerial institutions may be required, such as a pool of engineers, sociologists and economists to scrutinize the imported technology and advance innovation. These institutions are to address specific inter-disciplinary problems involved. These in turn have to be expressly linked to university research and risk capital to consulting engineering and manufacturing within the country itself as well as to corresponding institutions in the region where these developments are also taking place. National resources development planning, forestry, water, fisheries, etc., should be given due priority and be backed by practical actions because this is the kingpin in development.

The Mulingushi Reforms, Industrial Policies and Neglect for Science and Technological Development

The first move by the Zambian government to articulate an industrialization strategy was with the 1968 Economic Reforms announced by president Kaunda at the party National Council meeting at Mulungushi. The Mulungushi Economic Reforms, as they came to be known, were aimed at the state acquiring 51 percent shareholding in 26 privately owned companies. See Appendix 1.[4] These included both retail and manufacturing companies.

---

4. Kenneth Kaunda, *Zambia's Economic Revolution*, Mulungushi, April 19, 1968. Government Printers, Lusaka, pp. 38-43.

All were to come under the Industrial Development Corporation (INDECO). In addition to the establishment of this industrial structure was the strategy of developing industry through import substitution which was advocated by the First National Development Plan (FNDP).[5] This strategy sought to redress the situation of importing consumer goods by making them here in the country. Moreover, Zambia's reaction to the Unilateral Declaration of Independence (UDI) had been to produce locally those products which were being imported from Rhodesia and South Africa. To accomplish this, various policy instruments were used which included import tariffs, import licensing and exchange-rate controls through fixing the exchange rate and the allocation of foreign exchange. Let us examine them briefly before we look at the next section.

Import Tariffs

Import tariffs reflected government aspirations for industrialization by enforcing high tariff rates for consumer goods and luxuries and low rates for industrial goods. Intermediate and capital goods carried very low rates, even zero. This system influenced the industrial growth a lot till 1975 when it became overshadowed by the import licensing and foreign exchange allocation systems. Much as the system helped in the phenomenal rise in manufacturing industry, there were a few misgivings. The system encouraged a high degree of import content since industrial resource inputs were favored. It drained the spirit of development of capital goods and to some extent the utilization of local raw materials. This industrialization strategy also concentrated upon last-stage processing. The ILO mission observed that the industrial sectors import content increased from 39 percent in 1970 to about 60 percent in 1978.[6] This also meant a high degree of capitalization of the Zambian economy. Recently a National Cooperation Assessment and Programming mission restated this view when they observed that Zambian

---

5.   Republic of Zambia, *First National Development Plan 1966-70*, Lusaka, 1966.

6.   *Report of the National Cooperation Assessment and Programing Mission to Zambia*, New York, 1986, p. 91.

industries have tended to utilize large sophisticated methods of production. Smaller and labor-intensive options have not generally been used even where available. For example, in the manufacture of leather, cooking oil, maize grinding and bricks. These policy problems persist in our economy today.

Import Licensing

The Industrial Development Act 1977 provided that every manufacturer, small or big, had to license his or her enterprise. No clear-cut guidelines for giving licenses were laid down but all was done under the discretion of the Minister of Finance. This system of licensing was used as a control in foreign exchange allocation, i.e., every importer required a license from the Ministry of Industry and Commerce and Foreign Exchange from the Bank of Zambia. The consequences of these policies on the development, and utilization of local raw materials, small industries as well as results of R&D work have been in the reverse. No innovative spirit in industry and Government existed; instead dependence grew deeper. The years that followed from 1975 have showed that the industrial sector continues to exhibit a high degree of import dependence, poor maintenance of the productive assets and low capacity utilization. All these point to one thing; i.e., lack of capacity to adapt and innovate our industrial sector and maintain it on a self-sustaining basis. The success in development will be based on the ability to innovate in the manufacturing sector, expand existing output and services, and provide skills and technology where needed in order for us to properly exploit production opportunities. For this to be done, a support institutional framework explicit on how to go about this is needed. This also caused the neglect to develop a consistent science and technology policy which would have been the base of our industrial development.

Concluding Remarks

Zambia, like a large majority of African and Third World Countries, has paid little attention to the development of science and technology as a key sector for overall development in the contemporary. Instead, many such countries have given the science and technology sector only words-of-declaration plans or paper-and-policy speeches. The net effect has been nondevelopment of such a sector and consequently the dependence of the whole economy on externally generated scientific and technological knowledge. Their words of good intent are drowned out by the little financial commitments they accord to the sector compared to the social, political and even entertainment expenditures. In some cases, like Zambia, even a clear-cut institutional set-up to promote contemporary science and technology could not be constructed until almost 25 years after independence. The utilization of R&D results and the actual development of R&D is almost nil because investment incentives favor importation of capital goods and raw materials which have supported a relatively high-technology manufacturing industrial system specializing in last-stage assembling and production with little value added. The manufacturing industries have a seemingly indifferent attitude towards the financing of R&D. The critical notion in development that directs development through the utilization of existing resources, based on indigenously evolved and appropriately attuned scientific and technological endeavors, are very much wanting not only on the continent but in Zambia particularly, which is used as a case study in this chapter.

APPENDIX I

Companies Affected by 1968 Reforms

1. ARNOS INDUSTRIES

2. Monarch (Zambia) Ltd. (window and door manufacturers)

3. Crital-Hope (Zambia) Ltd.

4. Angolo-African Glass Company

5. P. G. Timbers

6. Baldwins Limited

7. Steel Supplies of Zambia Limited

8. Zamtimbia Limited

9. Johnston and Fletcher Limited (build material mMerchants)

10. Nicholas Quarries

11. Gerry's Quarries

12. Greystone Quarry (quarries supplying crushed stone)

13. Northern Breweries

14. Heinmuch's Syndicate (breweries)

15. Central African Road Services (transportation)

16. C.B.C. Stores

17. O.K. Bazaars

18. Standard Trading (rental trade)

19. Solanki Brothers Limited

20. Mwaiseni Stores Limited

21. Zambezi Sawmills

22. Mining Timbers (sawmills)

23. Zambia Newspaper Limited (newspaper publishing)

24. Irvin and Johnson Limited (fish marketing)

APPENDIX 2

Development of Research Centers in Zambia

	Research Centres	Year of Establishment	Field of Activity
1.	Meteorology Dept.	-	Weather forecasting problems of tropical meteorology.
2.	Geological Survey Department	1950	Geological, geophysical and geochemical mapping and pro-specting.
3.	Department of R&D (Roan Consolidated Mines Ltd., and Nchanga Consolidated Copper Mines Ltd.)	1954	Metallurgical mineralogical and chemical research and development work associated with metalliferous mining.
4.	Mines Safety Dept. (government)	1957	Particularly physics and ventilation studies.
5.	Mindex Exploration	1955	Geological, geophysical and geochemical exploration techniques for minerals.
6.	Hydrological and Hydrogeological branch (Dept. of Water Affairs	1950/52	Hydrological and hydrogeological observation and analysis of data.
7.	Department of Agriculture Research Branch - Mt. Makulu Research Center	1952	Crop research, soil research, irrigation studies and animal husbandry research.

8.	Central Veterinary Research Station	1928	Animal health and tsetse research.

National Council for Scientific Research 1969

9.	Food technology research unit	1969	Food technology research.
10.	Tree improvement research center	1965	Genetics, physiology and pathology of exotic and indigenous trees.
11.	Animal productivity	1963	Livestock reproductivity research.
12.	Pest research unit	1968	Tsetse and tick research.
13.	Water resources research unit	1966	Hydrological and hydrogeological investigations, water resources inventory compilation.
14.	Materials testing unit	1968	Materials testing services.
15.	Technical Services Unit	1973	Instrument servicing and repair, electronic and mechanical fabrication work.
16.	Radioisotopes Research Unit	1971	Application of radioisotopes.
17.	Building Research Unit	1969	Building materials & ceramic product development, design work for low-cost housing and other buildings such as rural schools and clinics.

18.	Cartographic and location analysis and printing unit	1980	Cartographic & quantitative locational analysis of data relevant to additional development planning.
19.	Industrial Minerals Research	1980	Investigate local industrial minerals & promote their industrial uses.
20.	Energy research	1980	Generating energy for domestic & industrial operations.
21.	Natural products research unit	1982	Investigation into plants with medicinal and pharmaceutical value.

University of Zambia 1966

22.	School of Medicine	1967	Varied medical & health research.
23.	School of Mines	1973	Geological, mineralogical metallurgical investigations.
24.	Technology dev. and advisory unit	1975	Dev. of design for equipment and tools suitable for small-scale industries.
25.	School of Engineering	1967/68	Various investigations, mechanical electrical & coil engineering.

26.	School of Natural Sciences	1966	Various investigations, chemistry, biology, physics and math.
27.	School of Agriculture	1971	Various investigations, crop breeding, animal breeding and husbandry.

APPENDIX 3

Examples of Scientific Development

Technologies Released to Industry

	Process/Product	Year	Company
1.	Soft drinks-tip-top (guava, lemon, pineapple)	1985	Copperbelt Bottling Company
2.	Masuku wine	1986/87	National Breweries
3.	High Protein Biscuits	1982	Diary Produce Board
4.	Cassava Biscuits	1984	With a private company

Development of Food Technology and Processes

Raw Material	Product	State of Develop.	Entrepren'rs
1. Soybeans	a) High protein biscuits	Transferred to Indus Dairy Produce Board (DPP)	Institutions for school-children and malnourished children.
	b) Nutrifex	Discussions with NFNC entered with malnour-ished children at UTH Paediatrics section undertaken with promising results	National Food & Nutrition Commission (NFNC).
2. Ground-nuts	a) Formula A	Complete technical know-how available with FTRU	ROP, VIS, SIDO and INDECO.
	b) Groundnut Oil		Refined Oil Products (ROP), Village Industry Service (VIS) Small-scale Industries Development Organiza-tion (SIDO).

3. Cassava	a) Biscuits	Complete technical	Mr. M. Sata Kafue Textiles Co.
	b) Starch	Complete technical know-how	Mr. Sata lira), Mbilishi (Lusaka), Nat'l. Drug Co. & Kafue Textiles interested.
4. Slaughter house waste from abortions	Gelatine	Complete technical know-how available	Cold Storage Board of Zambia.
5. Mollases	a) Industrial alcohol	Bench level	Nakambala Sugar Re-finaries
	b) Portable alcohol	Bench level	
6. Bananas	Strained baby foods	Bench level	Specialty Foods Co.
7. Masuku	a) Squash	Pilot community level	Dept. of Agricul-ture, NFNC, VIS, SIDO
	b) Wine	Pilot	Duncan, Gilbery & Metherson.
8. Pine-apple	a) Wine	Bench	Duncan, Gilbery & Metherson.
	b) Clarified Juice	Bench	Copperbelt Bottling Co.
9. Tomato	a) Juice	Pilot	Zamhort.
	b) Puree	Pilot	Copper-harvest.
	c) Paste	Pilot	Specialty Foods.

	d) Concentrate	Pilot	Zambia Bottlers.
10. Guavas	a) Concentrate	Pilot	Zambia Bottlers.
	b) Carbonated juice	Pilot	Copperbelt Bottling Co., Cadbury Scheppes.
11. Vegetables	Solar Dried	Bench	VIS, SIDO, NFNC and Ministry of Agriculture.

Chapter X

Obstacles in the Development of Science and Technology in Contemporary Nigeria

Julius Ihonbvere

The political economy of technology in the contemporary era cannot be properly comprehended unless it is contextualized within the historical experience of the relevant social formation. Such an approach would enable us to understand not only the specificity of the social formation but also how the institutions, structures, interests, forces, coalitions and contradictions generated by the historical experience have militated against or promoted by development of science and technology. In addition, the approach which enables us to consider the salience of history, power, production and exchange relations would also enable us to give particular attention to questions of state and class (social) relations as they affect technology in all its ramifications.

In the case of Nigeria, which is the primary concern of this chapter, the contact between the hitherto dynamic and developing though pre-capitalist social formation with the forces of Western imperialism have fundamental and monumental consequences which have continued to affect contemporary developments. Indeed, contemporary crisis, contradictions, conflicts, coalitions and problems which combine at various levels to reproduce Nigeria's backwardness in science and technology in the present can only be traced to this early contact with imperialism.[1] It is therefore necessary to pay particular attention to:

1. the nature of social relations prior to contact with imperialism;

2. the dynamic character of science and technology development in pre-capitalist Nigeria;

3. the nature of social classes created and nurtured by imperialism;

4. the specific character of the post-colonial capitalist state and the role it plays in production and exchange relations;

5. the accumulative base of the dominant classes and their relations with transnational capital;

6. the nature and solidity of inherited institutions particularly in the educational sector; and

7. the nature of Nigeria's location and role in the international division of labor.

If we put the factors above into serious consideration, it becomes obvious that issues of science and technology in Nigeria can only be considered as part of the ongoing class struggle which started from the colonial period. Ultimately, the direction of technology development would be expected to reflect the existing relations of power, production, exchange and accumulation.

Historical Experience, Science and Technology in Nigeria

Nigeria was colonized by the British. The period of colonization witnessed the massive exploitation of the country and the transfer of huge surpluses to the metropole. British colonialism culminated in the distortion, disarticulation and underdevelopment of the Nigerian social formation. Since the British were not interested in the overall growth and development of the Nigerian economy, issues of science and technology development were hardly considered as part of their priorities. Where it was found necessary to support the goals of domination, exploitation and surplus extraction and transfer, the British colonists reinforced superstition, ignorance and the reproduction of backward ideas in society.

Thus colonialism can be said to have laid the basis for science and technology retrogression in Nigeria. All autonomous possibilities for growth, development and self-reliance were terminated. New values, interests, relations of production and exchange, tastes, contradictions and institutions were introduced and superimposed on existing pre-capitalist conditions.[2] The economy was structurally incorporated into the periphery of world capitalism. As Yusuf Bangura et al. have noted:

> The effect of...imperialist onslaught and domination of the Nigerian economy was to subject the colonial economy to the dictates of the general worldwide capitalist system of development and crises; the economy lost its autonomous capacity to generate its own internal dynamic....Colonial capitalism was not interested in developing the productive forces in Nigeria but in producing primary commodities and a market for the goods of British and European companies.[3]

With the goals of colonial capitalism identified above, we can understand why the emphasis was on cash crop production for export and why it took the colonial state such a long time to become interested in spreading Western education beyond the limited horizons of the missionaries. Even then, the content of its educational agenda had very little to do with the development of science and technology. The principal interest was to produce personnel who can service the exploitative and repressive colonial machine - court clerks, interpreters, drivers, gardeners, policemen, messengers, produce buyers, clerks in the ministries, teachers, preachers, etc.

It is in the light of our understanding of the nature and implications of colonial domination that we agree with Aime Césaire that the colonial enterprise was

> ...neither evangelization, nor a philanthropic enterprise, or a desire to push back the frontiers of ignorance, disease and tyranny, nor a project undertaken for the greater glory of God, nor an attempt to extend the rule of law. To admit once and for all without flinching at the consequences that the decisive actors here are the adventurer and the pirate, the wholesale grocer and the ship owner, the gold digger and the merchant, appetite and force, and behind them, the baleful projected shadow of a form of civilization which, at a certain point in its history, finds itself obliged, for internal reasons, to extend to a world scale the competition of its antagonistic economies.[4]

Thus, Nigeria's historical experience was one of domination, degradation, exploitation, distortion, disarticulation and peripheralization in the global system. As Okwukiba Nnoli has rightly noted,

> ...the loss of local control over the new production process marked the beginning of the loss of self-reliance by the Nigerian economy. This loss was exacerbated and reinforced by the external orientation of the colonial economy. In sharp contrast to their pre-colonial production of goods and services for their own use, Nigerian participants in the colonial economy produced excessively for the world market.[5]

The process of this participation in the world economy albeit as a periphery and under conditions of extreme vulnerability to external manipulation, penetration and pressures meant that the dynamics of internal production for exchange could hardly generate a viable capacity for the development of science and technology. In fact, internal production and exchange processes were under the control of profit and hegemony seeking transnational corporations with little commitment to the Nigerian economy.[6] In sum, therefore, we can conclude this section by noting that the colonial experience left Nigeria a very unstable, poverty-stricken, foreign-dominated and vulnerable economy.[7] The inherited state structure at political independence in October 1960 was weak, unstable, non-hegemonic and with limited legitimacy. The dominant classes were dependent, corrupt and unproductive.[8] In any case the majority of them worked in the service of foreign transnational interests incorporated in Nigeria. The institutions and structures bequeathed to the "independent" nation were not only inefficient and ineffective, but were also alien to the objective aspirations and existential conditions of the people. The army, the police, tax agencies, marketing boards, bureaucracy and the court system were perceived as repressive and exploitative structures by the majority of the people.

Colonialism also left behind an air of distrust, ethnic and religious conflicts, instability and uncertainty. The edges of class antagonism had been sharpened and the politics of divide-and-rule on which the colonial state thrived had created not just regional imbalances but also regional antagonism. Limited social facilities had been concentrated in a few urban centers, infrastructures were poor, industrialization had been discouraged and the concentration on "each" crop production for export had affected the level of food production. Finally, the colonial educational system discouraged innovation, originality, and creativity, thus encouraging superstition and the pervasiveness of "colonial mentality" or "client mentality" - a mentality which discriminates against indigenous systems, products and ideals while glorifying anything foreign. This situation was strengthened on foreign aid, technology, information, investment and tastes as well as the marginalization of the economy in a highly competitive, exploitative and unequal world capitalist system.

Obstacles to the Development of Science and Technology in Nigeria

From our brief discussion thus far it is clear that we have identified five major obstacles to the development of science and technology in contemporary Nigeria. First, is Nigeria's contact with the forces of western imperialism. The consequences of this contact have already been highlighted. Second is the character and accumulative base of Nigeria's dominant classes which directly militates against the development of science and technology. The third obstacle is the nature of the state, particularly its inability to effectively mediate class contradictions and struggles, effectively plan for growth and development and strive for relative self-reliance. The fourth is the general condition of poverty, disease, unemployment, hunger, ignorance, superstition and other manifestations of under-development. These conditions reinforce the reproduction of science and technology backwardness. Finally, fifth is the degree of dependents domination, foreign exploitation, transfer of surpluses and marginalization in global capitalism.

To be sure, there are other issues we have left out. We believe however that these are the major obstacles to the development of science and technology in contemporary Nigeria. In any case, these are the roots of contemporary crises, contradictions and conflicts which result in the intensification of class struggles. The reproduction of contradictions in themselves militate against possibilities for technological innovation, absorption, replication and development.[9] Thus, the ability to mobilize human and material resources in support of the quest for science and technology development is often subverted by deepening contradictions and class struggles. Let us now look at the five major obstacles one after the other.

We have already highlighted the effects of colonial penetration, domination, exploitation and marginalization of the Nigerian economy. Suffice it to add that the colonialism, the social formation, lost indigenous and autonomous capacity to grow and develop. It lost the capacity to utilize existing resources for self-reliant development. The worldview, tastes aspirations and institutions implanted in the colonial period were such as to reinforce the country's backwardness. In all sectors of economy and society it is difficult to point at any concrete effort to push back the frontiers of poverty, disease, ignorance and superstition. The schools focused on classical languages, the arts and religion. Local creative capacity and local science were discouraged and discriminated against. Local products were branded illicit, illegal, inferior or dangerous. A case in point is the colonial attack on the production and consumption of local gin. The local potters, blacksmiths, carvers, goldsmiths, glass-makers, etc., were harassed and taxed out of operation or forced into cash crop production in order to earn the new colonial currency. The whole process of long-distance trade, communal relations and aspects of society which would have evolved into viable bastions of science and technology development were either stultified or destroyed.[10]

Perhaps a more critical obstacle to the development of science and technology since independence in 1960 is the nature or character of the Nigerian dominant class.[11] In particular, the accumulative base of this dominant class reveals a lot about its disposition towards the development of science and technology. It should be recalled that the Nigerian bourgeoisie is a creation of colonial capitalism. It was specifically created and nurtured to service the colonial capitalist economy in conjunction with imperialist capital. Through administrative, fiscal, political and other mechanisms, the colonial state ensured that the Nigerian bourgeoisie, irrespective of its fractional or factional position, remained a junior participant in the local economy.[12]

Thus, the colonial state ensured through direct and/or indirect policies the deformation and underdevelopment of the local bourgeoisie. The bourgeois class nurtured in the era of divide-and-rule politics as well as the politics of region, religion and ethnicity. What was more, the bourgeoisie was brought up to accept and believe in a worldview determined from Europe and America. In addition, the marginalization of the local bourgeoisie in the internal production and exchange processes meant that it was inherently unable to develop and implement concrete public policies likely to promote science and technology and thus move the nation towards self-reliant and self-sustaining growth and development. On the contrary, particularly after World War II, foreign capital embarked on a process of incorporating the emerging elites as legal advisers, partners, shareholders, political consultants, major representatives, etc. These were in themselves lucrative but dependent positions. Taken together, the positions produced a few wealthy persons but did not encourage the development of science and technology. Traders, importers and exporters do not need sophisticated technology to operate - ships could be bought and warehouses constructed at various trading centers.

The corruption of the Nigerian bourgeoisie which has reached legendary proportions is more a reflection of its accumulative base. In other words, because the bourgeoisie is largely divorced from production and accumulates through distribution and the direct looting of public funds, it hardly sees the need to pursue a serious science and technology program.[13] Its location in the distributive sector of the economy enables it to accumulate cash, build luxurious mansions, buy expensive electronic gadgets, throw scandalous parties and live a lifestyle completely alien to the imagination of the majority of Nigerians. The irresponsibility of this bourgeois class which makes it more interested in living for the present is evident in its failure to mediate class contradictions, provide effective leadership and initiate popular programs in support of overall national development. In contemporary Nigeria, in the areas of air travel, health services, education, electricity supply, communications, and security, the Nigerian bourgeoisie has created private alternatives to publicly sponsored services.[14] It is as if the bourgeoisie itself has lost faith in its ability to run the state for the good of its class, not to talk of the good of society.

Politics as practiced by the Nigerian bourgeoisie is a business; an investment. It has little or nothing to do with mass mobilization or public education. It is, on the contrary, a normless political struggle in which ethnicity, religion, state of origin, money, lies and "juju" are employed without restraint in the competition for public office. The bureaucracy, police, army, admission to schools at all levels, etc., are heavily politicised. Merit is easily sacrificed on the altar of expediency and other particular interests. In sum, we argue that the subservient, corrupt, unproductive, dependent and largely unpatriotic and irresponsible nature of the Nigerian bourgeoisie constitutes a major obstacle to the development of science and technology in Nigeria. Technology is perceived as the continuous importation of expensive mechanical and electronic gadgets as well as military hardware. Ability to manipulate, service and replicate the imported materials are hardly considered: they can always be replaced even with foreign loans. Ake is quite clear on the character of the Nigerian bourgeoisie:

> The indiscipline and anarchy within the Nigerian capitalist class has played a major role in reducing the prospects of capitalist development and deepening class contradiction and also jeopardizing the class interest of the capitalists. It has, behind the obscenely inflated contracts, the non-performance of contractors, and the massive scale of corruption which has reached new heights with the oil boom. It has contributed in no small measure to the serious balance of payments problems which the country is currently experiencing and the failure of important aspects in Nigeria's industrial and commercial strategies - in particular the promotion of industrialization....[15]

Ake's position, above, agrees with that of Eddie Madunagu who argues that the Nigerian bourgeoisie "are extremely irrational and underdeveloped" because "each Nigerian bourgeois is concerned with the promotion of his or her immediate exclusive interest."[16]

Another major obstacle to the development of science and technology is the Nigerian State. Like the bourgeoisie, the Nigerian State is a creation of colonial capitalism.[17] Pre-colonial state structures had been stultified, distorted and incorporated into a highly repressive, exploitative and violent state machine. Violence was used without restraint. The primary goal of the colonial state was the subjugation of the Nigerian people to the dictates of the metropolitan state and the massive extraction and repatriation of surplus. At independence, therefore, this state was seen by the Nigerian people as alien, repressive, exploitative and oppressive. Its public programs were not respected or acceptable to the people. It lacked the legitimacy required to mobilize the masses and inspire them to new levels of creativity and productivity. In addition, the post-colonial state was hardly autonomous as it was critically immersed in the class struggle on the side of the bourgeoisie. Consequently, the post-colonial state has had to contend with challenges from non-bourgeois forces on all fronts. Scarce resources are wasted on the acquisition of military hardware in order to contain the challenges. Because its policies commanded little acceptability, and its custodians commanded very limited legitimacy, the state has been unable to encourage its citizens and institutions to seriously pursue a credible and sustainable science and technology program. What is even more important is the fact that the state is visibly immersed in the ongoing class struggle on the side of the dominant classes. Its policies and actions are more often than not aimed at strengthening the dominant classes and imperialist interest while fragmenting, exploiting, repressing and domesticating non-bourgeois forces. Since political independence, its development plans, the recycling of oil rents, political pronouncements and more recently the implementation of the structural adjustment program have impoverished the masses of the people. Thus, to the vast majority of Nigerians, the state is not just exploitative and biased in favor of the rich, but it has also become largely "irrelevant."[18] Irrelevant to the extent that the state is very absent in a positive sense, in their daily struggles to reproduce themselves.

It is obvious that this is not the kind of state structure that promoted the industrialization and development of Japan, Russia, America and Germany. It is not the type of state structure that fires that level of patriotism and sacrifice in the citizenry that they constantly preoccupy themselves with how to control their environment and improve their society. The Nigerian state therefore is a direct obstacle to the development of science and technology not just through its inconsistent and sometimes confused policies but also through its commitment to the reproduction of a neo-colonial educational system and collusion with imperialist interests to explicit non-bourgeois forces. The brain drain is only a reflection of the inability of the state and its custodians to maintain a conducive environment for its best brains to flourish.

The general conditions of want, poverty, disease, hunger, illiteracy and insecurity constitute direct impediments to the development of science and technology. When people are poor and have to devote their energies to the struggle for survival, their horizon and worldview would certainly be stunted.[19] As well, a poor, hungry and illiterate populace has little time to study developments in other societies, innovate and improve on pre-existing rudimentary technology. The Nigerian society, in spite of several development plans, an oil boom and the allocation of billions of Naira to poverty-alleviating programs, is still plagued with want and poverty. Unemployment, inflation, prostitution and other features of backwardness still exist. Poverty breeds and reproduces superstition and ignorance. Poverty breeds vulnerability to manipulation and commitment to the status quo. The International Labor Office was quite clear on the extent of poverty in Nigeria when it noted in its 1981 report that:

> There is still much acute poverty in Nigeria....Since the overall rate of economic growth has apparently been fast in the past two decades, there should be by now signs of substantial improvements in the living conditions of the majority of the Nigerian people. These are not evident...large numbers are worse off in numerous respects especially in the rural areas but also in the big new slums of the cities.[20]

The general inefficiency and ineffectiveness of public institutions, the poor state of social services, absence of communication facilities and the backwardness of the school system all reflect the state of underdevelopment. The deepening crisis of the economy compelled the leaders recently to impose a World Bank and IMF-sponsored Structural Adjustment Program (SAP). Since the SAP was imposed on a sea of poverty, disease, inequality and alienation, the program has worsened existing problems culminating in riots. As retired General Olusegun Obasanjo, Nigeria's former head of state noted in a recent review of the SAP: the program was promoting foreign penetration and massive transfer of surplus; a serious brain drain; alienation; and a drastic reduction "in the living standard of all classes of productive workers."[21]

Our position is that the existence of poverty, especially within the context of an exploitative state and class structure, works against the development of science and technology in Nigeria. Since serious poverty-alleviating strategies are not in place, it means that the masses of the people would continue to divert their energies and scarce resources to the search for food, drugs, jobs, transportation, accommodation and other pressing problems. They would have very limited time to reflect, think, study their problems, understand the achievements of other societies and make contributions to the development of science and technology.

To be sure, the obstacles discussed thus far need to be contextualised within the global environment within which Nigeria operates. As a former colony, the Nigerian economy is inextricably linked with those of the metropolitan capitalist powers. Given the internal underdevelopment and structured incorporation of the social formation into the world system, Nigeria has, since the colonial period, become dependent on external aid, technology, high-level manpower, information and military hardware. Profit- and hegemony-seeking transitional corporations dominate all sectors of the economy and block possibilities for science and technology development. This is in any case the most assured way to maintain and reproduce Nigeria's science and technology dependence. Internal conditions of instability, poverty, and weakness of state and class forces contribute to the reproduction of international technological dependence. As Clive Thomas has noted:

> ...technological dependence is an expression of both the internal social condition of the under-developed countries, as well as the product of certain exploiting relations between the under-developed countries and the main capitalist centers. From the international standpoint, therefore, the issues of technological dependence and/or cooperation..., have to be posed in the context of two very broad and general class struggles.[22]

Nigeria not only lacks the relevant data base required to support the continuous development of science and technology but the bourgeoisie has initiated and nurtured a process where the technological innovation of earlier centuries has come to be replaced by "technological importation." As victims of imperialist manipulation the bourgeoisie has swallowed hook, line and sinker the nation's erroneous view that technology can be transferred to Nigeria once the "international behavior" is right. In other words, once transnational investors are given a free hand to operate internally and the capitalist West is given all possible diplomatic support, technology, now defined as complex computers, military hardware and sophisticated machines would be transferred to Nigeria. The Nigerian state and its custodians believe and accept this situation in spite of the obvious fact that the country has at the moment no credible domestic producer goods sector to stimulate local productive forces. Thus, a material base for technological innovation is kept permanently repressed. The informal sector of artisans and other self-employed craftsmen who constituted the nucleus of earlier technologies are discriminated against and harassed. The practitioners are branded and their workshops are demolished at the slightest excuse.

Finally, because of the largely unpatriotic and corrupt disposition of the Nigerian bourgeoisie, it has come to develop a dangerous consumption attitude. The agents of the state travel around the world making excessively generous concessions to foreign investors and importing all sorts of fancy gadgets. They overlook internal obstacles to science and technology development while reinforcing internal structures which reproduce external dependence. The Nigerian dominant class overlooks the fact that the complex interplay of internal and external forces is required to promote the processes of technology innovation, absorption, replication and transfer where possible. Alexander King is quite clear on this point:

> Why then has technology transfer not been achieved to a much more significant extent? The answer seems...to be not so much in the inadequacies of the transfer system as such, the machination of transnational enterprises or iniquities in the terms of transfer, but in a failure on the part of both donors and receivers to appreciate that successful transfer is a somewhat complicated socio-economic process with many variables and social conditions beyond the mere introduction of packets of knowledge and technical know-how.[23]

The Nigerian elite believes that science and technology can be encouraged through the proliferating of Colleges of Science and Technology, polytechnics and Universities of Science and Technology. It is bad enough that most of these institutions of science and technology run courses in law, the humanities and social sciences. The schools are poorly staffed, lack adequate research and teaching facilities and contribute very little to the quest for science and technology development. Though the number of students in such science-biased institutions have increased tremendously in the past decade or so, Segun Osoba has argued that:

> ...in spite of this apparently phenomenal growth in the provision of facilities and student places for technical education, the Nigerian economy has continued to be plagued by severe shortage of qualified technical personnel in both its public and private sectors...(the) persistent shortage of qualified technical and professional manpower in a situation of escalating growth in technological education is rooted deeply in the structures of the political economy of Nigeria; and it is not capable of being resolved, as the Government wrongly believed by purely technical measures like training more people....[24]

Part of the explanation for this rather dismal situation is because:

> ...the curricula operated in most secondary schools and in virtually all primary schools have not been planned to produce young school leavers with adequate scientific knowledge and consciousness to enable them to benefit from higher technical and scientific education...It is this very weak scientific content in the curricula of the lower levels of the educational system that is primarily responsible for the unsatisfactory situation in our polytechnics which government official policy statements constantly decry but which the authorities and their planners do little or nothing about.[25]

To be sure, the poor development of science curricula in primary and secondary schools generates a sort of bias against the study of science technology subjects. In fact, most students who pursue courses in medicine, pharmacy and engineering do so not for any intrinsic interest in science and technology but because in neo-colonial Nigeria these courses are adjudged to be very lucrative.[26] Taken together, science and technology policies do not in any way reflect the objective conditions and needs of society. At the educational level, the disjuncture between the existential needs of Nigerians are hardly considered, science and technology courses and programs are heavy on theory and very weak on practicals. This weak empirical or practical base of Nigeria's science and technology program in the contemporary era cannot be divorced from the corrupt and generally misguided planning perspectives of the Nigerian state and its custodians. As Segun Osoba has rightly noted:

> One of the most disastrous consequences of this kind of irrational planning is the divorce of technological education in Nigeria from the real life of the people and the real needs of the society. For instance, there is the unrelatedness of a great deal of our technological training programs to the employment requirements of a rapidly changing society. This results in critical shortages of appropriately qualified technical manpower in several sectors of the economy (i.e. petro-chemicals, aviation, irrigation, metallurgy and materials engineering) at the same time as there is serious unemployment among graduates of universities and technical institutes in fields like engineering physics, town planning and construction related disciplines...[27]

It is obvious therefore that the obstacles to science and technology development in contemporary Nigeria are fundamentally class based.

CONCLUSION

In this chapter we have only tried to provide a short summary to critical issues raised by other contributors to this volume. We have looked at the obstacles to science and technology development in contemporary Nigeria from two broad but complementary perspectives. The first is the country's historical experience which culminated in colonialism, domination, exploitation, distortion, disarticulation and incorporation into an exploitative world capitalist system as a periphery. Science and technology development in the colony was only a reflection of the needs of metropolitan capitalism. The institutions, interests and social relations introduced at this point ensured the overall backwardness, underdevelopment and dependence of neo-colonial Nigeria.

The second fundamental obstacle we identified was the post-colonial interplay of class forces. Here we argued that the neo-colonial inheritance, the culture of dependent accumulation, conditions of foreign domination as well as the nature and character of the state, social classes (especially the bourgeoisie) and politics combine at varying levels to militate against science and technology development in contemporary Nigeria. Nigeria's dominant class is largely comprador and rentier in character.[28] Its accumulative base is largely in lucrative but unproductive sectors of the economy with limited requirement for science and technology. Politics and the control of state power, or at least access to the state, are major avenues for primitive capital accumulation through sinecure appointments, corruption, consultancies, and inflated contracts. The net result of accumulation without involvement in production is the direct neglect of science and technology.

It would appear therefore, that the solution to the present crisis cannot be understood outside an overall restructuring and transformation of the social formation. The state of science and technology development in all societies reflects the state of development of productive forces and the overall nature of politics, power, production, exchange and consumption. Insofar as Nigeria remains a dependent, dominated and disarticulated society and the dominant classes remain unproductive and corrupt and the state remains unstable and non-autonomous, science and technology will remain stunted.

NOTES

1.   See Toyin Falola and Julius C. Ihonvbere (eds.), *Nigeria and the International Capitalist System*, Denver, Lynn Reiner for GSIS, 1988.

2.   See Toyin Falola (ed.), *Britain and Nigeria: Exploitation and Development?* London, Zed Press, 1985.

3.   Yusuf Bangura, Rauf Mustapha and Saidu Adamu, "The Deepening Economic Crisis and its Political Implications," in Siddique Mohammed and Tony Edch (eds.), *Nigeria: A Republic in Ruins* Zaria, Department of Political Science, ABU, 1986, p. 173.

4.   Aime Cesaire, *Discourse on Colonialism* New York, Monthly Review, 1972, pp. 10-11.

5.   Okwukiba Nnoli, "A Short History of Nigerian Underdevelopment," in his collection, *Path to Nigerian Development,* Dakar, Codesria, 1981, p. 165.

6.   See *ibid.* and works by Segun Osoba, Tim Shaw, Yusuf Bangura, Bala Usman, Claude Ake, Eme Ekekwe and Bade Onimode.

7.   See Julius Ihonvbere (ed.), *Dependent Capitalism and Crisis in Nigeria,* Benin-City, Jodah Publishers, forthcoming.

8.   See Eme Ekekwe, *Class and State in Nigeria,* London, Longman 1985 and Eddie Madunagu, *Problems of Socialism: The Nigerian Challenge,* London Zed, 1982.

9.   See Julius O. Ihonvbere and Toyin Falola, "Technology Transfer to the Third World: Obscurantism, Myth and Implications," *Journal of General Studies* 5 & 6 (1), 1984-85 and Segun Osoba, "Political Economy of Technological Education: Myth and Reality," *African Historian,* Vol. XI, 1983/84.

10.   See Falola (ed.), *Britain and Nigeria, op. cit.*

11.   See Claude Ake (ed.), *Political Economy of Nigeria,* London, Longman, 1985, Julius Ihonvbere and Toyin Falola (eds.), *State, Class Society and in Nigeria,* Ibadan, Heinemann, forthcoming and their *The Rise and Fall of Nigeria's Second Republic, 1979-84,* London Zed Press, 1985.

12.   See E. O. Akeredolu-Ale, *The Underdevelopment of Indigenous Underdevelopment in Nigeria,* Ibadan, Ibadan University Press, 1975 and Bade Onimode, *Imperialism and Underdevelopment in Nigeria,* London Zed Press, 1982.

13. Eddie Madunagu, *Nigeria: The Economy and the People-Political Economy of State Robbery and its Popular Democratic Negation*, London New Beacon, 1983.

14. Julius O. Ihonvbere, "The State and the Irrationality of the Bourgeoisie: An Examination of how the Nigerian Bourgeoisie Subverts its own Future," paper presented at a conference on "Awolowo: The End of an Era?" organized by the Obafemi Awolowo University Press, Ile-Ife, October, 1987 and Segun Osoba, "The Deepening Crisis of the Nigerian National Bourgeoisie" *Review of African Political Economy* (13), 1978.

15. Claude Ake, "The Nigerian State: Antinomies of a Periphery Formation" in his edited Volume, *Political Economy of Nigeria, op. cit.*, p. 16.

16. Eddie Madunagu, *Problems of Socialism, op. cit.*

17. See Ihonvbere and Falola (eds.), *State, Class and Society in Nigeria* and Gavin Williams, *State and Society in Nigeria*, Idanre Afrografika, 1980.

18. For a detailed discussion of 'the irrelevant state' see Claude Ake, "Theoretical Notes on the National Question in Nigeria," mimeo Port Harcourt; and J. Ihonvbere and E. Ekekwe, "The Irrelevant State...in Nigeria,' in J. Ihonvbere (ed.), *Dependent Capitalism and Crime in Nigeria*, Benin-City Jodah Pub., forthcoming.

19. For details on the extent of poverty in Nigeria see ILO, *First Things First: Meeting of the Basic Needs of the People of Nigeria*, Addis Ababa, JASPA, 1981.

20. *Ibid.*, p. V.

21. Olusegun Obasanjo, *National Concord*, 12 Dec. 1987, p. 15.

22. Clive Thomas, "Scientific and Technological Cooperation: Reciprocity and/or Dependence," in *Scientific and Technological Innovation, Self-Reliance and Cooperation*, proceedings of the 8th International Conference of the Institute for International Cooperation, April 1976, Ottawa, University of Ottawa Press, 1977, p. 122.

23. Alexander King, "Use and Abuse of Science and Technology for Development," in A.J. Dolman and J. Van Ettinger (eds.), *Partners in Tomorrow: Strategies for a New International Order*, New York, E. P. Dutton, 1978.

24. Segun Osoba, "Political Economy of Technological Education..." *op. cit.*, pp. 7-8.

25. *Ibid.*, p. 10.

26.   To be sure professionals in these areas are not only more conservative and even reactionary but also appear to enjoy a higher standard of living even in periods of serious economic crisis.

27.   Segun Osoba, "Political Economy of Technological Education, p. 12.

28.   For a discussion of the rentier state in Nigeria see Toyin Falola and J. Ihonvbere, *The Rise and Fall of Nigeria's Second Republic, 1979-84*, London (ed.) Press, 1985.

Chapter XI
Historical Perspectives on Technical Cooperation in Africa

Carlos Lopes

Summary (in English)

According to the Berg Report the level of basic discomfort in African countries has reduced considerably since Independence. This is presumably so in health, education, and the overall growth of institutions, a view which contradicts more pessimistic perspectives on development trends in Africa in these areas. We must however emphasize that there has been in Africa the perpetuation of the models and systems of dependence created on the eve of Independence, a trend which becomes more visible as time goes on. Unfortunately economic analysis has paid very little attention to such issues. In this chapter the point is made that there has been in Africa a reproduction of a model of knowledge transfer which is specific to this region, a fact which explains the poor record of technical cooperation in Africa. The culture-centered approach claims that the reasons for 'African backwardness' lie in the realm of culture. We point out however, that the perpetual elusiveness of technical cooperation is not due to an alleged cultural specificity but rather to a specificity that is structural, and which goes back historically to the colonial model of accumulation, a model preoccupied with the exploitation of natural resources, and the phenomenon of unequal exchange. Subsequent developments have simply helped to consolidate Africa's dependence, and have been less concerned with genuine cooperation.

UNE PERSPECTIVE HISTORIQUE DE LA COOPERATION TECHNIQUE EN AFRIQUE

par Carlos Lopes

Il est généralement admis qu'il est difficile de cerner un sujet sans le situer dans son contexte et, selon la nature du thème, l'aspect historique peut avoir plus au moins d'importance. Pour ce qui est de la coopération technique il est évident que nous avons besoin du recul historique pour des raisons qui sont liées aux innombrables interprétations que l'on prête aux termes "coopération" et "technique," mais aussi aux différents concepts sous-jacents à cette conjonction de mots.

Nous avons essayé de passer en revue les principales étapes de l'évolution de l'idée et de la notion da coopération technique, et pour cela nous sommes trouvés obligés de remonter jusqu'aux origines de ce phénomène: la période coloniale. Par la suite nous avons analysé l'integration de l'Afrique dans l'économie mondiale et la construction des justifications idéologiques permettant la consolidation de diverses formes de depéndance.

1.   Le Modele Colonial

La Conférence de Berlin de 1884 à surtout servi à la délimitation de l'accès au fleuve Congo, grande porte ouverte pour l'hinterland africain. C'était aussi l'occasion de définir les principes qui allaient régir la croisade coloniale. Dorénavant les rapports économiques inégaux qui supportaient les relations entre africains et européens vont passer à un stade supérieur, celui de l'occupation territoriale et de la subordination politique. Ce virage, annoncé par Berlin, va se concrétiser dans les fameuses campagnes de "pacification," qui serviront à délimiter les frontières sur la base du principe de l'occupation territoriale.

La révolution industrielle dans certains pays européens avait démontré le besoin de débouchés et de relais commerciaux dans le monde entier et cette nouvelle étape du développement capitaliste était imprégnée d'illusions coloniales. Dans cette fin du XIXe siècla rien de plus normal que de penser que l'Afrique potantiellement pouvaient "servir," autant que les Amériques ou l'Inde, d'autant plus que ses richesses minérales étaient déjà connues.

Le modèle d'accumulation coloniale de l'époque était peu sophistiqué: exploiter les ressources naturelles, par des moyens qui mettaient à profit une main d'oeuvre non qualifiée, abondante et soumise à un rapport de forces très particulier. Ce modèle commençait déjà a négliger la possibilité d'utilisation de l'Afrique comme marché pour certains produits manufacturés européens, créant les mécanismes d'un commerce très limité aux intérêts de certains agents économiques, qui resteront très périphériques dans les préocupations des métropoles.

Interdépendence et échange inégal

Il est évident que la base du modèle colonial est la nature pré-capitaliste des rapports de production des sociétés africaines, qui sont restées liees à une exploitation de la terre par des systèmes de lignage, avec des rendements très bas. Ces formes traditionnelles d'organisation économique engendreront des analyses dualistes qu'il serait fastideux de présenter ici. Retenons toutefois que l'interdépendance entre les différents modèles économiques, endogènes et exogènes, va en s'accentuant, augmentant l'échange inégal entre les métropoles européennes et leurs territoires d'outre-mer.

A ce titre il nous semble intéressant d'illustrer cette interprétation par un extrait d'un discours prononcé par M. Brevie, à l'ouverture de la session du Conseil de Gouvernement de l'AOF, le 14 décembre 1931,[1] sur l'interdépendance des territoires:

> "J'ai donc pensé que le moment était venu de demander à la Métropole de relever l'Afrique occidentale d'un effort qu'elle ne peut plus matériellement soutenir et de pendre en main une cause qui est aussi la sienne, tant du point de vue politique que du point de vue économique et social, puisque toute diminution du pouvoir d'achat de la Fédération aggrave la situation déjà difficile de l'industrie et du commerce français dont nous sommes à la fois les fournisseurs et les clients."

Et il ajoute qu'il n'y a pas "un commerce métropolitain et un commerce colonial," mais plutôt un commerce français" parfaitement formé et ferme."

En effet M. Brevie se tourne vers les autorités centrales qu'il accuse, entre les lignes, d'oublier ses possessions africaines au moment ou le monde vit une grave récession, qui conduira, d'ailleurs, à la deuxième querre mondiale. Il essaye de mettre en exergue le besoin de palier aux difficultés de la métropole par un programme d'*aide*:

> "Il m'a paru que le moment était opportun pour reparler d'organiser une aide de la production coloniale à la Métropole, d'instaurer une politique cohérente et suivie, qui permettrait à la France de mettre son domaine d'outre-mer en pleine valeur et de se présenter dans la lutte économique mondiale forte de toutes les richesses que lui apportent ses enfants de Métropole et des Pays d'outre-mer."[2]

Cette *aide économique* que les colonies pouvaient apporter à la France était néanmoins conditionnée par une amélioration des rendements de ses économies dépendantes. Pour M. Brevie cela pouvait se faire à condition qu'une *aide technique* soit mise à la disposition des administrations des territoires d'outre-mer. A titre d'exemple il considère que:

---

[1] In *"Annuaire de Documentation Colonials Comparée - Afrique Occidentale Française,"* Bruxelles, 1931, pp. 373-380.

[2] Idem, p. 385.

> "...le GovernE Général a mis à l'étude toute une série de
> questions intéressant la productivité générale de la Fédération
> (AOF) visant particulièrement à améliorer, c'est-à-dire à
> rendre plus économiques les transports (...); la constitution
> entre les mains de l'indigène d'un outillage agricole susceptible
> d'augmenter le rendement de son travail (...); l'adoption d'un
> texte relatif à l'organisation du crédit agricole (...)."[3]

M. Brevie insiste sur le caractère "technique très marqué" de l'aide qu'il préconise.

Il est facile de se rendre compte que nous touchons à la racine de l'approche assistance ou coopération technique.

L'assistance technique "avant la lettre"

L'apport, la transmission ou l'échange de connaissances techniques a une origine très lointaine. Alexandre le Grand avait envoyé des missions en Chine pour qu'elles apprennent l'art de tisser la sole. A l'époque de la Renaissance il y avait beaucoup d'échanges d'informations techniques et pendant la nuit coloniale le commerce atlantique a sevi de relais à la diffusion des techniques d'extraction et utilisation du fer en Afrique occidentale, etc.

Ce n'est pourtant que tout récemment que nous entendons parler de ce transfert comme étant de l'assistance. Ce sont les oeuvres missionnaires qui utilisent cette expression en premier lieu. Accompagnant la fixation territoriale des colons, les missions catholiques, protestantes et d'autres égilses réformistes américaines, s'attacheront à diffuser des connaissances considérées comme *fondamentales* pour l'amélioration des conditions de vie des populations africaines.

Les missions créent des écoles, des hôpitaux et centres de santé et des domaines agricoles. Pour les administrations coloniales leurs préoccupations ont un fondement religieux mais sont utiles économiquement: il s'agissait bel et bien d'augmenter les rendements et de transmettre les habitudes économiques exogènes, qui étaient, nécessairement, celles du capitalisme occidental.

---

[3] Idem, p. 383, oestrid.

Ces oeuvres religieuses auront bientôt pour compagnie des sociétés philanthropiques - tels les Fondations Ford et Rockefeller, des Etats Unis - qui tenteront de diffuser des connaissances scientifiques et techniques déjà considérées comme *appropriées aux sociétés indigènes*.

Cet effort sera consolidé par les administrations coloniales, à travers des plans de développement, qui intègrent ou associent les initiatives des oeuvres religieuses ou sociétés philanthropiques.

La Grande-Bretagne a créé son programme d'aide à travers une loi de 1929[4] qui prévoyait l'octroi d'un million de livres sterling par an aux propositions présentées par les autorités des colonies au Colonial Office. Ce montant a été porté an 1940 - en pleine guerre - à 5 millions par an et, pour la période 1946-1956, un total de 120 millions de livres a été dépensé.

L'Afrique recevait à peu près 60% de cette somme, et (par ailleurs) consacrait 50% de ses financements aux domaines de l'éducation, de la formation de fonctionnaires coloniaux et de la recherche. Ce domaine était considéré comme prioritaire méritant même, pour les pays du Moyan Orient, la création en 1946 d'un organisme spécifique pour les actions d'assistance technique, le Middle East Office.[5]

L'autre grande puissance coloniale, la France a créé, en 1946, le FIDES (Fonda d'Investissement pour le Développement Economique et Sociale)[6] dont le but était "l'établissement, le financement et l'exécution de plans d'équipment et de développement des territoires d'outre-mer." C'est d'ailleurs grâce à ce Fonds que Marginot lança son plan de construction de routes et ports dont le but était de créer un seul marché de Dunquerque à l'équateur, afin que les colonies aident l'économie de la Métropole.

Plus que dans le modèle britanique l'école va devenir l'affaire de l'administration territoriale: faire accepter l'idéologie du colon, sa lanque et sa culture comme moyens d'atteindre la civilisation. Dans le cas français

---

[4] Voir "L'Assistance Technique aux pays insuffisamment développés" 1ère partie, *La Documentation Française,* n. 1928, 1954, p. 4.

[5] Idem, p. 4.

[6] Ibidem, et Jacques Adda/Marie-Claude Smouts, *"La France face au Sud, le miroir brisé,"* Karthala, Paris, 1989, pp. 27-35.

aussi une grande importance est donnée à la recherche scientifique, avec la création de nombreux instituts et centres, tous encadrés par un Conseil Supérieur de la Recherche Scientifique et Technique d'Outre-Mer.

Les belges et les portugais, ces derniers à partir de l'Acte colonial approuvé en 1930, s'attacheront aussi à l'idée d'une assistance à leurs administrations coloniales pour qu'elles développent des programmes de formation des élites et fonctionnaires coloniaux indigènes, nécessaires pour mieux encadrer les populations et améliorer leur productivité.

Ces initiatives que nous venons de présenter constituent les traits communs entre le modèle colonial et celui qui le succèdera. Les administrations coloniales et métropolitaines établissent une relation qui sera reproduite, par la suite, entre l'Etat indépendant et le donateur. Et au centre de cette relation nous identifions le concept d'aide ou d'assistance technique.

Les concepts d'aide et de rentabilité

Tout d'abord précisons que l'idée d'aider un pays ou un territoire est chronologiquement postérieure a une utilisation du terme dans une base plus générale, individuelle ou collective (aider le prochain), de connotation religieuse.

L'identification territoriale, quand on parle d'aide, est donc des plus importantes. Elle est en plus accompagnée de justifications géopolitiques (possessions, colonies, pays sous tutelle) ou économiques (rentabilité).

L'aide est d'abord perçue comme de l'assistance. L'idée d'assistance est interprétée comme le salut de l'assisté, ce qui implique l'existence de deux pôles qui ne sont plus régis par les règles de la solidarité sociale ou de la générosité personnelle mais plutôt par la responsabilité sociale d'un groupe. D'après Rahnema "even when assistance is not prompted by the fear of the poor as a danger to society, or by the donor's individual salvation, the finality is the same: the greater moral and material comfort of the intervenor, rather than an existential necessity for the latter to understand the poors' predicament and to respond to their sufferings. The assisted is, in the last analysis, an object, an instrument and a means to a different end."[7]

L'utilisation de l'expression *assistance technique* n'est donc pas un hasard. Toujours d'après cette interprétation il s'agirait du transfert des connaissances dans un sens, dans une direction, en fonction de la richesse technique des uns par rapport à la pauvreté technique des autres. Et tout cela dans puel but?

Dans le but d'augmenter les capacités productives - les rendements...- de présider à la définition des modèles, de pénétrer et développer les marchés, d'étendre la sphère d'influence, bref de contrôler la reproduction économique d'une société, d'un territoire.

Pour cela il faut instrumentaliser l'aide afin qu'elle puisse garantir une rentabilité.

2. L'Impact des Deux Guerees Mondiales Dans Les Programmes D'Assistance

La recherche de laa paix au lendemain des deux querres mondiales commande tous les sentiments. La dévastation et les destructions provoquées par le conflit se font sentir surtout en Europe. Epargnée territorialement par la guerre l'économie américaine va assurer le rôle de locomotive de la croissance capitaliste. C'est donc tout naturellement qu'à la fin des années 50 les regards se tournent vers Etats Unis.

Le discours unitaire et universel de la Charte des Nations Unies, approuvée à San Francisco en 1945, ne doit pourtant pas faire oublier que nous sommes en présence d'un monde ruiné, avec une Europe particulièrement divisée, avec des tensions et rivalités de tous bords, et avec des idéologies en confrontation croissante.

Les grandes puissances issues de la 2ème grande guerre - les Etats Unis et l'URSS - semblent s'intéresser davantage à la reconstruction

---

[7] Majid Rahnema, *"Global Poverty, a pauperizing myth,"* mimeo, s/d,s/l, p. 10. L'auteur cite Georg Simmel, pour lequel "assistance to the poor holds, in legal teleology, the same position as the protection of animals." Ce qui l'amène à conclure que "modernized poverty, defined by assistance, constitutes a radical break from other vernacular relationships based on gift, charity, or even protection (p. 11).

européenne. L'Afrique est encore perçue comme l'arrière-cour des puissances coloniales européennes.

Les initiatives des Etats-Unis d'Amérique

Chacune des grandes puissances s'était assuré une zone d'influence géostratégique et politique en Europe. La guerre froide garantira l'étanchéité de cette division en Europe occidentale et orientale, avec des régimes politiques très différents.

Les Etats-Unis décident, au lendemain du cauchemar hitlérien, de lancer un programme d'aide a l'Europe occidentale, à grande échelle, le fameux Plan Marshall.

Ce Plan était assis sur un certain nombre de présupposés qu'il importe de rappeler:

-aider la reconstruction d'une Europe qui doit résister au communisme;
-démontrer la générosité instinctive du peuple américain;
-remettre en état des sociétés modernes, hautement productives, avec des ressources humaines très qualifiées, temporairement endommagées par la guerre;
-positionner les Etats-Unis comme leader incontestable de l'économie mondiale.

Les principes directeurs du Plan Marshall, avec les volets humanitaire, défensif et économique, seront par la suite utilisés dans l'ensemble des programmes d'aide des Etats Unis. L'ambiguïté résultant de l'inexistence de deux éléments primordiaux du Plan, dans les programmes d'aide aux autres régions du globe - le principe de remise en état et le flux important de capitaux assuré par le Plan (plus de 3% du PNB) - sera dépassé par le Président Harry Truman, avec les propositions du "Point IV."[8]

---

[8] Dane un discours du 20 janvier 1949 le Président Truman a présenté son programme de reconstruction mondiale et son 4ème thème-connu depuis comme le "Point IV"-prévoyait l'adoption d'un programme d'assistance technique aux pays sous-développés. Dans ce discours le Président Truman passe en revue tous les aspects relatife à la mise en place d'un programme élargi d'assistance technique, avec les caractéristiques qu'aura le programme des Nations Unies, qui est né à la même époque. Un des arguments les plus utilisés par le Président est celui de l'intérêt d'une telle assistance pour les pays riches: "...nous avons souvent grand besoin de leurs ressources (celles des pays sous-développés) et du produit de leur labeur. Si la productivité et le pouvoir d'achat de ces pays s'accroissent, notre propre industrie

Le "Point IV" est la suite logique d'un certain nombre d'initiatives américaines pour institutionnaliser les mécanismes de l'aide fournie par leur administration. Ainsi le Président Roosevelt avait déjà créé dans les annes 30 l'Interdepartmental Committee on Scientific and Cultural Cooperation et l'Institute of Inter-American Affairs, tous les deux dépendant du département d'Etat et chargés de fournir à l'Amérique Latine et aux Caraïbes des spécialistes, du personnel scientifique et de procéder à des échanges d'étudiants. D'autres départments de l'administration américaine, tels celui du Commerce, de l'Agriculture ou de la Santé avaient aussi envoyé des missions et des spécialistes dans d'autres régions, notamment en Egypte et au Liberia.

C'est cette expérience qui sera formalisée et consolidée par le "Point IV," qui a créé des grands espoire dans les pays déshérités, espoirs non totalement confirmés parce que le Congrès n'a pas suivi les idées de son Président avec l'enthousiasme qu'il attendait. Ceci a d'ailleurs contribué au développement d'une argumentation de moins en moins humanitaire et de plus en plus géostratégique. Il est donc possible de dire que dès son départ le "Point IV" sera aussi le modèle de l'aide politisée, assurant les intérêts de celui qui donne plutôt que les désirs de celui qui reçoit.

L'adaptation du modèle colonial

Les puissances coloniales ne sont pas restées insensibles à cette nouvelle approche des relations internationales et ont essayé d'adapter à leur manière les principes du "Point IV."

Pour des pays comme la Grande-Bretagne, la France, la Belgique et le Portugal il s'agissait de consolider les mécanismes d'aide déjà existants entre les métropoles et leurs colonies, notamment africaines.

Un an après le discours du Président Truman sur le "Point IV" la Conférence des Ministres des Affaires Etrangères du Commonwealth a décidé de lancer le Plan Colombo pour le développement de l'Asie du Sud et

---

et notre propre agriculture en bénéficieront...En augmentant la production et le revenu national des pays économiquement moins développés nous ajoutons à notre propre stabilité économique," (extraits du discours du Président Truman, in "L'Assistance Technique aux pays insuffisamment développés," 3ème partie, *La Documentation Française*, n. 1930, 1954, p. 3.

du Sud-Est. Ce Plan sera la première initiative d'envergure du Commonwealth dans le domaine de l'assistance technique, et il est intérsseant de constater que, malgré la participation de pays non affiliés au Commonwealth (dont les Etats Unis) l'ensemble des échanges préconisés par le Plan se feront à partir des territoires de l'ancien empre britannique.

La France, elle, se déploie en Afrique. Au développement du FIDES[9] il faudra désormais ajouter le renforcement des échanges économiques entre les territoires, et le renforcement d'une identité francophone. Les mécanismes qui consolideront ces tendances sont le Programme Cadre, qui sevira de base à l'accession à l'indépendance des territoires africains, et à la création de la zone franc.

"L'assistance technique (française) revêt essentiellement les formes suivantes: a)-envoi d'experts à l'étranger, b)-réception de boursiers et stagiaires, c)-visites en France de personnalités étrangères, d)-diffusion de documentation technique (publications scientifiques, ouvrages techniques, catalogues industriels) et éventuellement envoi de matériel d'études et de recherches."[10]

Obligée, plus que la Grande Bretagne, de se replier sur l'Afrique, la France regarde avec préocupation les premiers mouvements d'émancipation secouant les populations. La politique d'intégration économique de ses colonies, en train de devenir indépentantes, coûte, à la fin des années 50, à peu près 1.5% de son PNB, chiffre jamais égalé depuis par la coopération française."[11]

Contestée dans certains milieux[12] l'aide française, rebaptisée "coopération" au lendemain des indépendances, sera l'instrument de

---

[9] Fonds d'Investissements et de Développement économique et Social, créé en 1946 dans le but d'élaborer des plans de modernisation et d'équipment de l'Union Française.

[10] "L'assistance...," 3ème partie, p. 10.

[11] *"La France...,"* p. 29.

[12] Un livre polémique publié en 1963 (Edouard Bonnefous, *"Les milliards qui s'envolent, l'aide française aux pays sous-développés,"* Fayard Paris) a provoqué une réaction colérique du Ministère de la Coopération à travers une brochure intitulée *"Non le France ne gaspille pas ses millards"* (Paris, 1965), où on pouvait lire ceci, "Loin d'être dilapidés, notre aide est infiniment

perpétuation du modèle colonial. A tel point qu'il faudra attendre 1981 pour que l'aide publique française soit officiellement évaluée sans les DOM-TOM.[13]

Signalona aussi que les puissances coloniales (Grande Bretagne, France, Belgique, Portugal) et l'Union Sud-Africaine et al Rodhésie du Sud, décident, aussi en 1950, d'établir une Commission de Coopération Technique en Afrique aud du Sahara, la CCCTASS. La justification présentée, lors de la création de cetted Commission à Paris, était le besoin de renforcer l'échange d'information sur des domaines d'intérêt commun, et de consolider des actions concertées qui se faisaient sans un cadre juridique précis. La Commission en viendra à créer un Secrétariat à Londres, en 1952, et établira 5 bureaux spécialisés dans des domaines précis. Cette Commission aura des relations privilégiées avec certains organismes des Nations Unies, comme la FAO at OMS.

Cette Commission contribuait au renforcement des liens entre les administrations territoriales, et après les indépendances le cadra qu'elles ont établi servira de base au recrutement d'experts sur lea problèmes africains, dans les domaines techniques.[14]

Le développement du multilatéralisme

La Charte des Nations Unies a donné un nouveau souffle aux intentions de la SDN (Société des Nations), dans le domaine de la coopération entre les Etats. Elle était le fondement de base pour le développement du multilatéralisme.

Le précurseur multilatéral de la coopération technique est le Comité pour la Coopération Intellectuelle, créé par la SDN en 1922, et qui donnera

---

plus judicieuse. Si au depart, il y a su des dons, et si, parfois, des gaspillages ont pu être signailés, il est à signaler que nous nous sommes efforcés très vite de lier les crédits accordés au point qu'en 1964, ceux-ci représentaient 60% du total et que cette tendance ira en se renforçant au cours des années à venir," in "La France...," p. 31.

[13] Departments et Territoires d'Outre-mer. Voir "La France...," p. 30.

[14] Voir "L'Assistance...", 2ème partie, p. 16.

plus tard naissance à l'UNESCO.[15] Ce sont les intentions de la SDN, consolidées par la Charte de l'ONU et le modèle américain du "Point IV" qui seront à la base de la création en juillet 1950 du PEAT (Programme Elargbi d'Assistance Technique).

Dès le départ les experts des Nations Unies ont dû débattre de deux questions essentielles pour bâtir le système de l'aide:

a) - adapter ou restructurer les mécanismes existants - avec d'un côté ceux qui défendaient un greffage sur l'expérience occidentale de relations avec les anciens pays coloniaux, et de l'autre côté ceux qui préconisaient des réformes de structures;

b) - séparer l'aide financière de la coopération technique- puisque les institutions issues de la Conférence de Bretton Woods (Banque Internationale pour la Reconstruction et le Développement et Fonda Monétaire International) s'occupaient du premier aspect, il fallait bâtir le PEAT comme complément aux interventions de la BIRD et du FMI.

Ces débats ont façonné le PEAT, et son successeur le PNUD. Les deux auront un mandat limité a la coopération technique et les deux hériteront des concepts développés dans les années 50.[16] Le Technical Assistance Board, organisme chargé d'exécutar le PEAT mettra en place toute la structure que nous connaissons de nos jours: les bureaux résidents, la relation entre organisme de financement central (PEAT, PNUD) et les agences d'exécution du système des Nations Unies (tels l'OIT, UNESCO, OMS, FAO, ICAO, etc), l'approche projet, le rôle des experts....Et ses activités vont croître à une vitesse de croisière.

Une évaluation du l'PEAT en 1965 constatait qu'il avait assuré l'envoi de 32,000 Hommes/années d'experts at 32,000 bourses d'études, sans

---

[15] A ce propos il est intéressant de constater qu'en 1923 un des membres de ce Comité, Albert Einstein, adresse au Secrétariat de la SDN une lettre précisant que "Dans ces derniers temps, je suis arrivé à la conviction que la SDN n'a ni la force ni la bonne volonté pour remplir sa grande tâche," et il précise en conséquence "je vous pris de rayer mon nom de la liste des membres de la Commission" (SDN docs C.129.1922..XII et c. 312.M.155.1922.XII, Bibliothèque Dag Hammerskjold des Nations Union, New York).

[16] Pour une histoire détaillés de l'EPTA voir "*15 years and 150,000 skills, and anniversary review of the United Nations Expanded Programme of Technical Assistance,*" prepared by the Technical Assistance Board, United Nations, New York, 1965.

compter les centaines de séminaires, voyages d'études et autres activités de formation sur le tas, estimant le nombre de bénéficiaires à 150,000, répartis pour 130 pays et territoires. Il est intéressant de constater qu'au départ les récipiendaires ont toujours préféré l'utilisation d'experts aux bourses d'études ou autres formes de coopération technique. Ces experts venaient pour la plupart de l'Europe (58.3%) ou des Amériques (25.1%), l'Asie (11.3%), le Moyen Orient (3.7%) et l'Afrique (1.7%) n'ayant pas de poids significatif.[17] L'envoi d'experts a consommé 60% des financements, tandis que les activites de formation se limitaient à 14%.[18]

L'Afrique a utilisé 20.3% des ressources du PEAT jusqu'à 1965, mais sa part va passer de la dernière à la première place pendant ces premières 15 années d'existence du PEAT. L'importance accordée au continent ne cessera d'augmenter au fur et à mesure que les pays africains deviendront indépendants.

Normalement le TAB organisait une mission exploratoire - une des premières missions est allée en Libye pour faire une évaluation des besoins - qui recommandait les secteurs dans lesquels le PEAT devait intervenir. La priorité était donnée à l'industrie et à l'agriculture. Ces missions intégrant les spécialistes des agences du système des Nations Unies, devaient tenir compte d'un certain nombre d'éléments:

a) - promotion de l'économie du pays (Théorie des avantages comparatifs);
b) - le gouvernement devait "demander" l'assistance technique;
c) - le gouvernement récipiendaire devait contribuer au projet;
d) - l'assistance ne devait pas interférer dans les choix politiques et économiques du gouvernement;
e) - les experts devaient connaitre les aspects culturals du pays auquel ils seraient envoyés;
f) - l'existence et disponibilité d'homologues.[19]

---

[17] "15 years...," p. 32.

[18] Idem, p. 31.

[19] Ibidem, p. 9.

Tours ces éléments sont devenue des piliers des formes multilatérales de coopération technique et il n'est pas difficile de constater que si les intentions étaient nobles, le fonctionnement de ces principes a toujours constitué un problème.

En 1954 le gouvernement français a proposé un changement dans les règles de fonctionnement des "Programmes de pays" (qui étaient et sont encore de nos jours la structure de base du système PNUD) pour que les sommes attribuées aux pays ne soient pas décidées d'avance par l'PEAT, mais plutôt en fonction des vrais besoins exprimés par les pays récipiendaires.   Ces "French proposals," timidement approuvées par la résolution 542.B.II.(XVIII) de l'ECOSOC et 831(IX) de l'Assemblée Générale, sont toujours restées lettre morte, faute d'une alternative au fonctionnement en vigueur.   Trente cinq années après, le débat introduit par les NATCAP ressemble beaucoup aux incertitudes de 1965, comme le démontre ce passage d'un document officiel:

> "There was a danger in this process that the result would tend to be a mere catalogue of projects to be undertaken in each country concerned, rather than a co-ordinated and meaningful programme truly dictated by the country's needs in their order of priority, in relation to one another and to economic imperatives and objectives...."[20]

Malgré ces échecs le modèle multilatéral des Nations Unies sera considéré, pendant un certain temps, comme plus sain que les coopérations bilatérales.   C'est ainsi qu'une multitude d'organisations gouvernementales verront le jour et s'inspireront des modalités développées par le PEAT et le PNUD.

3.   Theories Du Developpement

"Dans la plupart des cas, le sous-développement est vécu par des non-occidentaux.   Mais la plupart de ceux qui analysent les symptômes ou construisent des théories sur ses origines ou la façon de l'éliminer sont des occidentaux ou des intellectuels occidentalisés appartenant aux pays

---

[20] Ibidem, p. 12.

économiquement arriérés. C'est un des aspects les plus remarquables des études, aujourd'hui très nombreuses, qui sont consacrées au développement."[21]

Cette phrase de Tibor Mende a fait verser beaucoup d'encre au début des années 70, ou la mode était aux études du développement. De nombreux Instituts et centres de recherches, surtout en Europe, ont alors construit de nombreuses théories interprétatives du phénomène de sous-développement.

Aussi bien l'approche dualiste (double secteur traditionnel et moderne) celle des étapes (les stades de développement) que celle du développement inégal (centre et périphéries) adressent des critiques aux schémas de coopération technique existants. Mais toutes ces théories négligent l'analyse des aspects concrets liés au renforcement institutionnel, élément principal des arguments et des échecs du système en vigueur.

Nous pensons que le mérite de la perspective historique est celui de démontrer l'articulation chronologique entre les différentes initiatives de transfert de connaissances. A ce titre il est intéressant de voir l'impact des relations internationales sur le plan politique dans l'évolution du système.

Les effects pervers de Bandoeng et le dialogue Nord-Sud

Les mouvements indépendantistes et de libération nationale qui ont vu le jour au lendemain de la 2ème guerre mondiale ont radicalement changé le visage de l'Afrique et de l'Asie. Il était donc tout naturel de voir naître une solidarité afro-asiatique, officiellement ratifiée par la Conférence de Bandoeng, en avril 1955.

Le charismatique Président Sukarno, d'Indonésie, a pu constater la naissance d'un nouvel axe des relations internationales qui sera connu sous l'appellation "tiers-mondisme," et qui donnera naissance à de nombreuses organisations, dont celle qui convoquera la Conférence Tricontinentale de la Havane en janvier 1966: l'OSPAA (Organisation de Solidarité des Peuples d'Asie et d'Afrique) au sigle de laquelle on ajoutera, après la Conférence, un autre A, pour Amérique Latine.

---

[21] Tibor Menda, *De l'Aida a la recolonisation,"* Seuil, Paris, 1972 et 1975.

Le terma tiers-monde avait été proposé pour la première fois par Alfred Sauvy, dans l'hebdomadaire français "Observateur,"[22] en 1952. D'après lui "nous parlons volontiers des deux mondes en présence, de leur guerre possible, de leur coexistence, etc, oubliant trop souvent qu'il exists un troisième, le plus important et en somme le premier dans la chronologie. C'est l'ensemble de ceux qu'on appelle en style Nations Unies les pays sous-développés."[23]

Les idées tiers-mondistes auront un grand impact dans toutes les discussions au sein des Nations Union, et par la création du mouvement des non-alignés et le "Groupe des 77" les pays dits sous-développés réclameront une participation plus active dans la sphère internationale. Or il est èvident que les débats sur l'aide et la coopération - technuque ou autre - n'ont pas été épargnés.

La critique de l'aide trouvera des très nombreux supporters et l'acidité des arguments obligera une recification du langage utilisé. En fait il n'était plus possible d'admettre, comme principe de base, que les pays du tiers monde n'avaient pas les capacités intrinsèques pour le développement, mais, plutôt, que les moyens faisaient défaut. De la à considérer la coopération technique comme une forme subalterne d'assistance il n'y avait qu'un pas que beaucoup ont pu franchir sans crainte.

Les effets pervers de Bandoeng ont continué à se faire sentir jusqu'à la recherche illusoire d'un nouvel ordre économique international, projet pour lequel les nations du tiers monde se sont mobilisées dans la décade de 60. Un espace de dialogue a pu être institutionnalisé au sein des organisations internationales.

Des mesures correctives ont été proposées dans le cadre de ce dialogue Nord-Sud, notamment en ce qui concerne:

a) - L'assistance financière et technique aux pays pauvres-qui devait être intensifiés et dans la mesure du possible utilisée pour la partie la plus défavorisée des populations;

---

[22] Voir Yves Lacosts, *Unité et diversité du tiers monde, deux représentations planétaires aux stratégies aur le terrain,* La Découverte, Paris 1984.

[23] Idem, p. 14.

b) - Les cours des matières premières - qui devaient être stabilisés et les termas de l'échange améliorés;
c) - Le commerce international - qui devait permettre un accroissement significatif des exportations des pays du Sud en direction du Nord.

Tant qu'a duré la période d'expansion la plus longue qu'ait connu l'économie mondiale, ces plaidoyers semblaient raisonnables et ont été acceptés par les Nations Unies. Malheureusement tout ceci est resté lettre morte et les termes du dialogue Nord-Sud ont par la suite été répétés ad nauseeam.[24]

Ce qui amène Tibor Mende en 1972 aux interrogations suivantes:

"On commence seulement à poser les questions fondamentales, celles qui comptent réallement. Les dons, ou l'aide étrangère en général, peuvent-ils stimuler le développement? Les gouvernants qui ont reçu les dons étaient-ils ou non capables de les utiliser dans le but désiré? Les hypothèses de travail des économistes, des administrateurs et de tous les organismes officiels et officieux qui se sont occupés de d'aide reposaient-elles sur des faits d'expérience? Entre-temps la seule chose qui n'ait pas changé, c'est le problème lui-même.

La coopération technique sera placée au centre des débats peut-être parce que c'était le maillon faible de la chaine et il y aura une réduction importante de cette forme d'aide dans toutes les régions du monde à l'exception de l'Afrique.

4. Conclusion: L'essor de la Cooperation Technique

D'après le Rapport Berg[25] le niveau des contraintes fondamentales des pays africains a considérablement été réduit depuis les indépendances. Ceci est dû surtout aux efforts dans le domaine humain (santé, éducation, amélioration des capacités institutionnelles), et contrarie des analyses plus pessimistes sur les possibilités de développement de l'Afrique dans ce Domaine.

---

[24] *"De l'aide a le recolonisation,"* p. 6 et 7.

[25] World Bank, *"Accelerated Development in Sub-Saharan Africa,"* Washington, 1982.

Il faut, cependant, constater une perpétuation en Afrique subsaharienne des modèles et systèmes de dépendance forgés à la veille des indépendances. Ce fil conducteur devient plus visible avec une perspective historique. Malheureusement les analyses économiques font très peu de recours à ces dimensions.

Nous avons essayé de démontrer qu'an Afrique il y a une reporduction d'un modèle de transfert de connaissances qui est particulier à cette region. C'est ce fait qui explique le peu de résultats perspectifs de la coopération technique en Afrique.

Comme le prétend l'approche culturaliste - qui prétend qu'il existerait des raisons culturelles au "retard africain" - nous sommes d'accord qu'il y a bel et bien una spécificité africaine, qui expliquerait, en partie, un perpétuel essor de la coopération technique. La grande difficulté c'est que la spécificité n'est pas culturelle mais structurelle.

Ceci explique cela.

CONTRIBUTORS

Nurudeen Abubakar is a Research Fellow at the Centre for Nigerian Cultural Studies, Ahmadu Bello University, Zaria. He has specialised in the history of technology in the Northern Nigerian region with specific emphasis on metallurgy.

Bala Achi has published widely in several journals. He has been doing research on the military history of the Nigerian region. He is about to complete his doctorate.

Donald Chanda has published widely in the area of Science and Technology development in the context of the Zambian region. He has also written on general development issues. Donald Chanda lectures at the University of Zambia.

Aliyu Idrees has been preoccupied with reconstructing the history of the Nigerian Nupe-speaking region. Having lectured at the University of Ilorin, Nigeria he joined the staff of the Federal University of Abuja where he now lectures.

Ahmad Kani is the Dean of Arts at the Usman Dan Fodio University, Sokoto, Nigeria where he is a Reader/Associate Professor. He has published extensively.

Julius Ihonbvere lectured at Obafemi Awolowo University as well as the University of Port Harcourt. He has served as a Visiting Professor at the University of Toronto in the 1990/91 session. He "joined" the staff of the University of Texas in the 1991/92 session as an Associate Professor. Professor Ihonbvere is the author and co-author of several texts.

Carlos Lopes is an expert in Human Resources and Technical Cooperation with a background in academic and political institutions as this relates to African Development. He received his doctorate from the University of Paris and has published extensively. He has served as a member of the Experts Committee for Science and Technology of the U.N. Economic Commission for Africa, as Chariman of the Sub-Commission for Research, Training and Cultural Action of Angola, Cape Verde, Guinea Bissau and Mozambique and is now at the UNDP, New York. He comes from Guinea Bissau.

L. R. Molomo has engaged in research on aspects of Material Culture in the Zimbabwe region. He has lectured at Kenyatta University Kenya and is at the moment Head of Department of History, Hillside Teacher's College, Zimbabwe.

Olowo Ojoade lectured for several years at the University of Jos where he has served as a Dean of the Faculty of Arts, and a Director of the Centre for General Studies. His major area of specialisation is in the analysis of West African proverbs. Professor Ojoade is a member of several international organizations.

Gloria Thomas-Emeagwali has lectured at the Ahmadu Bello University, Zaria the Nigeria Defence Academy, Kaduna and the University of Ilorin, Nigeria. She has served as a Visiting Scholar at the University of the West Indies, and more recently Oxford University where she was a Visiting Fellow and Senior Associate Member. She is an Associate Professor at the Central Connecticut State University.

A. Zack-Williams lectures at Lancaster Polytechnic, Preston. He has lectured at Bayero University Kano, Nigeria and the University of Jos, Plateau State, Nigeria. He is a member of the editorial Committee of the Review of Political Economy, London. His major area of research is Sierra Leone and he has published widely on that region as well as on other parts of West Africa.

Tea

and

Ceremony

Experiencing Tranquility

DIANA SALTOON

INTRODUCTION BY
DENG MING-DAO

ILLUSTRATED BY
EVELYN HICKS

A skysociety™ West Coast Edition
See www.skysociety.com

Published by Robert Briggs Associates

Designed by Anita Jones

Library of Congress Control Number: 2004103387
ISBN: 0-931191-18-1

dedication

For my mother,
who believed in ritual
and always made a fine cup of tea.

Contents

Illustrations vii

Introduction ix

    Deng Ming-Dao xi

One: Tea and Tradition 1

  1 Why Tea, Why Ceremony? 3

  2 An Overview of Tea Cultures 13

  3 Varieties of Tea 43

  4 The Experience of Tranquility 59

Two: Practical Adaptations 87

  5 Ways to Tranquility 89

  6 Brewed Tea 99

  7 Whisked Tea 113

  8 Reflections 149

Glossary 153

Some Sources 175

Acknowledgments 187

Bibliography...189

Illustrations

Why Tea, Why Ceremony?

\# 1 Teacup and madeleine 4

\# 2 Nakamura Sensei 6

An Overview of Tea Cultures

\# 3 *Camellia sinensis* 14

\# 4 The whisk 18

\# 5 *Song* bowl: Temmoku type 25

\# 6 Russian with a samovar 36

Varieties of Tea

\# 7 Tea in a glass 49

The Experience of Tranquility

\# 8 Guest drinking tea 60

\# 9 Japanese tea house 67

\#10 Guest scooping water..74

\#11 Alcove with scroll and flowers 76

\#12 Lacquered stand with tea utensils 77

\#13 Tea bowl, cloth, whisk, and scoop 78

\#14 Kettle in hearth 84

Ways to Tranquility

\#15 *Gong fu* Chinese ceremony 102

\#16 An adaptation of English tea 110

\#17 A table-style whisked tea 124

\#18 Folding wiping cloth: parts 1 and 2 127

\#19 Folding wiping cloth: parts 3 and 4 127

\#20 Wiping caddy and tea scoop, 2-part 129

\#21 Drying tea bowl, 3-part 132

\#22 Whisking tea 133

\#23 Tea outdoors 147

\#24 Stone basin and water ladle 151

Introduction

by

Deng Ming-Dao

x

Introduction

Tea and Ceremony reveals the Way of Tea—*Chado*—as a living path. Like any true path, it is there for any person to take, without limits on who that person may be. Embarking on this path requires only sincerity and a willingness to understand that tea, in all its beautiful forms and with all our appreciations, can be a lifelong journey. This book begins by showing us tea's global history and then explains how that rich history can inform our own attraction to tea. More important, it reveals how we can then use tea in our daily lives for removal and spiritual perception.

Chado is intimately connected with Zen, and Zen in turn is part of a living culture. It cannot be separated from the world. That's good for two reasons. First, it means that its spirituality is an integral part of daily life. Second, it means that Zen can be entered from a variety of angles. Since it is an integral part of life, it cannot be alien to what we do each day. Since it is part of daily life, deep understanding in any number of activities can lead to spirituality. In the monasteries, the masters engage in many seemingly ordinary tasks—cooking, sweeping, repairing walls, gardening—all because they know that to declare some acts spiritual and others worldly is to make foolish distinctions.

Yet the work of a master can have a different quality than the work of someone who doesn't care about what they do. Those who regard work as drudgery stay separate from all that surrounds them. They just want to "get it over with," or they wish they were "anywhere but here." The master does the task to the fullest degree necessary, and doesn't wish to be anywhere but there.

So daily tasks may have spiritual potential, but there is evidently a difference in the *how* of doing it. In some Zen temples, for example, a student is told to do a certain housekeeping task but is not told how to do it. There is a reason for that, of course. The student must go out and talk to more experienced monks, and in so doing he must discover the method and meaning of the task. Not knowing how or what one must do to accomplish a certain goal may seem confusing, but it can also spur one to concentration and determination. Such focus is the beginning of the spiritual essence of what may seem to be an ordinary job.

At other times, the masters give us certain forms to follow. This is true in a variety of Asian traditions. For example, a painting student might spend ten years just painting plum flowers. A calligraphy student will copy the scrolls of masters repeatedly for years. A martial arts student is taught long series of movements derived from the battles of past champions. An ascetic may repeat the same mantra for decades on end. Even in legends and folk tales, plants, animals, and people will follow certain procedures for self-cultivation. In the famous Chinese story, "Journey to the West," many characters, including the central one, Monkey, practice for centuries in order to develop supernatural skill and seek enlightenment.

Viewed from this perspective, *Chado*, the Way of Tea, is not just a formal method of drinking tea. It is a practice, with spirituality in the details. The rustic tea room, the carefully chosen scroll, the tea bowl, the quality of the water, and the actions of the server are crucial. True, it takes many years to practice the full Way of Tea, but this means it is a long path with many stages that will hold our attention. A journey can be more delightful if we know it will last. Knowing that should not be a barrier to taking the first few steps to begin our enjoyment.

The art of tea exists not to make the drinking of a delightful beverage complicated, but to show us the path taken by previous masters. We can find the same vistas they themselves glimpsed. In this case, the vistas in a small tea room and the vistas in a bowl of tea are equal to the view from a holy mountain or the visions of a hermit-poet.

A bowl of tea contains the whole world. It is round, reminding us of the cycles of life, the shape of our planet, the return of each day in the sunrise and the clouds. The bowl is forged from the Five Elements–metal, wood, water, fire, and earth—all are necessary to make it. The bowl may have a rustic feel, but it is undeniably a made thing: a piece of art with a practical function created by another human being. Tea represents life and death. We pick the best leaves, we save them, their green as intense as all the days of spring, and we whisk them with living water heated over a living fire.

After, though, we must discard what is left (tea left overnight will develop mold), and so there is an ending to our experience, and a need to let go of what we had.

Breath is at the center of our experience. We cannot live without air. We cannot do anything—sing, talk, work—without using our breath. The masters of breath—*qi* in Chinese, *ki* in Japanese—tell us that all life is energy, and all energy is of the breath. It so happens that *Chado* is also a way of the breath. The tea hut is ideally in a garden, close to the dewy breaths of the morning leaves. The sweep of the broom, the exhale of the host as she enters the room, the steam rising from the kettle, the air whisked into the tea, the fragrance from the surface, the breath of those who drink from the bowl—all these are ways that our individual breathing falls into harmony with others, with the garden, and with the greater world on the breezes that flow over the garden walls. Our breath is shaped in very much the same way that the exercises of the meditation masters shape our inhalations and exhalations.

Tea is also healthy for us. It refreshes us, awakens us, and cleanses us. It neither sedates us as wine might, nor over stimulates us as coffee might. There are antioxidants in teas, and scientists are discovering even potential anti-cancer benefits. Tea heals, and healing has always been part of the meaning of spirituality.

Yet what of tea outside of Asia? If Zen and tea are united in Japanese culture, what does that mean for those of us in the United States and other countries in the West? The answer to that question has led to a great deal of experimentation and, in some cases, mistaken attempts to bring Zen and tea to the West. It's absurd for non-Japanese to try to mirror Japanese culture. That only creates separation. We will not find the spirituality that is integral with daily life as long as we are trying to copy someone else.

Equally perplexing is to Westernize *Chado*. There are too many conflicts, too many different assumptions. In the modifications, we inevitably have to discard the Japanese identity of *Chado* and we end up diminishing what we originally valued.

The only possible solution to this paradox is long study and gradual absorption of *Chado*. Only then will it become a natural part of our lives and we a natural part of it. Through such a gentle process, spirituality will come as a natural outcome of our activities. If we can approach *Chado* in this way, then we honor both it and ourselves.

Introduction

Tea and Ceremony leads us into this process. Diana Saltoon begins by explaining her early exposure to tea as a young girl—a perfect introduction, innocent, but containing the essential elements of offering and sharing. Her overview of tea culture is not simply a concise description of tea's history around the world, but it also shows that the wonders of tea as a beverage and an art are open to anyone around the globe. She then orients us to the authentic Way of Tea as it is practiced in Japan (a way she now teaches as a fully accredited teacher). Her "Practical Adaptations" section is where we can see ways in which *Chado* might be practiced in our own lives to bring us tranquility—even if we are used to tea in a mug or want to have a good drink of tea on a hiking trip.

There are two concepts here that are worth more discussion: ceremony and removal. A misunderstanding here would be tragic. It would mean that we might turn away from a wonderful art, fearing the dryness of rituals we might find meaningless. Similarly, a misunderstanding of removal might mean we miss the integration so important to rooting *Chado* in our own culture.

Ceremony must not be seen as valuable in itself. We are not looking for elaborate forms of etiquette or a way in which we might outperform other people in appearing cultivated. Unless ceremony can be seen as an art—perhaps even a choreography—we cannot enjoy it. It is a dance, a graceful and balanced set of movements that expresses our own thoughts and gives us access to the thoughts of the tea masters who created them. Like a dancer who appreciates the dance created by one of the teachers before her, engaging in ceremony means we give ourselves over to the wonders of making an art with our bodies, minds, and hearts. No intellectuality will help. No anxiety will make us more graceful. We need only do the ceremony to unlock its meaning, and in so doing we unlock self-cultivation.

Similarly, we must not see removal as meaning a removal from our lives. That would be separation, not the integration into the living way that is *Chado*. We are not trying to get into some exalted dimension. There is no life to be lived but this one. However, in giving ourselves over to art, we can create a wonderful confluence of the moment we find ourselves in, the tea we take into our bodies, the sharing with others, and the appreciation of nature. That is a deeper, more mindful moment. If there is any removal, then, it is removal of the obstacles to tranquility.

Tea and Ceremony is an important book, the culmination of decades of study. The author has faced the same obstacles you and I have faced. She has had to balance her interest in tea with marriage, family, and work. Nevertheless, she has long pursued both Zen and tea, and her recognition as an experienced teacher is evidence of her accomplishment. Her message for all of us is that *Chado* is a way to be walked even as we are making our other necessary pursuits. It isn't necessary to leave this world to find tranquility. The author shows us how a simple bowl of tea brings us its own peace.

When one goes to see a teacher, it is common to bring a gift. When one goes to have tea with someone, a present might also be in order. Here, though, the teacher is giving us a lovely opening into the world of tea. If you accept that, just as you would accept a bowl of tea from her hands, the magic, the wonder, and the tranquility of this centuries-old practice can be yours. *Tea and Ceremony*, then, is Diana Saltoon's beautiful and precious gift to us all.

Deng Ming-Dao
San Francisco, California
March, 2004

One

Tea and Tradition

1

Why Tea, Why Ceremony?

My relationship with tea began in early childhood. I was born in Singapore, then a British colony and home to an extraordinary mix of Chinese, Malays, Indians, Singhalese, and Europeans. Most of them drank a great deal of tea. As a little girl, I was as familiar with tea as a child growing up in the United States might be with milk. But tea was only one of the thousand delights of my childhood, a commonplace among the tempting curries, noodles, spiced rice, mangos, and *chendol*—a shaved-ice ambrosia of coconut cream, palm sugar, and green pandanus jellies offered at the fabulous food stalls along the city's streets and alleys.

When I was seven, a Chinese opera set up a huge tent on a vacant lot in our neighborhood. Clanging cymbals lured me across the street, where I wandered among stalls and dressing nooks in which the actors were preparing for a performance. Dazzled by their sequins, I felt lost among the blue, red, gold, and yellow-painted faces until a kindly woman who tended costumes beckoned me. When I approached, she offered me a tiny handleless cup filled with tea, which she had carefully poured from a small ceramic teapot. It was not at all like our tame home-brewed tea, but fragrant and slightly bitter. It was from China, exotic, unknown, and it was my introduction to tea as something other than my customary childhood fare, something more alluring.

In the late 1950s I would have tea at the old Raffles Hotel on Singapore's Beach Road. There, under the gentle whirl of the ceiling fans, afternoon tea was served in porcelain cups, accompanied by tiny sandwiches and an array of cakes and eclairs. Being grown-up by then, I would wear my best dress and enjoy conversations with friends and visitors. More than once my mind floated with the piano music out from the lounge and over the open-air Palm Court. I did not yet know that by taking tea and letting myself be transported beyond the sweeping staircases and the stuccoed columns, I was seeking a momentary removal from the problems and routine of every-day life.

Figure 1

Teacup and madeleine

Years later, I read in *Remembrance of Things Past* how a cup of tea and one of those squat, plump little cakes called "petites madeleines" had caused Proust to "shudder" as he recalled a "powerful joy." I remembered those afternoons at the Raffles, but it was not until the 1980s that I was able to make the connection between tea and that joy. I was living in San Francisco and had begun to practice *chanoyu*, the Japanese tea ceremony, and to appreciate the vision of Lu Yu, China's patron saint of tea, and the wisdom of Sen Rikyu, Japan's greatest tea master. Until then, I had been simply one of the many millions of people who had developed an enduring affection for this age-old beverage.

My experience of tea eventually merged with a profound interest in Zen Buddhism. Both enhanced my life, providing relief through the trials of marriage and divorce as well as insight into the confusion and hopes encountered in daily life. In 1992, early one morning in London (where I was traveling on business), I sat down on a bench in Holland Park to watch some ducks floating in a quiet pond. The harmony of movement and stillness produced a tranquility of mind and body both refreshing and lucid.

It occurred to me that the Japanese, who had interned my family and me in Singapore during the war, maintained a culture which in later life offered me so much freedom through the practice of Zen meditation and the Way of Tea. I came to see that tea and Zen complemented each other and were aspects of a single practice, one that has helped me understand how and why, at the beginning of the twenty-first century, meditation and ceremony are more important than ever. I have found that my Zen decision to be in the moment can be attained in the making of a bowl of tea.

When patiently and attentively prepared, a bowl or cup of tea becomes a means to achieve calm, much like meditation. However, while meditation requires the body to be stilled and the mind emptied, the tea ceremony calls for an awareness of place and procedure. It clears the mind, opens the heart, and establishes an affinity between host and guest. It might be said that whereas meditation is passive and silent, the tea ceremony is active and sensitive to sound and sight and, above all, to the dignity of human exchange.

I first encountered traditional Japanese tea culture in 1986, when I was invited to a tea ceremony lesson at Green Gulch, the Zen center just north of San Francisco. The occasion was fortuitous, for I met Soshi Nakamura, who later became my teacher, or *sensei*. Even though I had read about the tea ceremony, I was quite unprepared for what I witnessed. As I sat in the tea room and observed the lesson, the commitment of students and teacher impressed me, and I began to see how and why the ancient ritual of tea was so exquisitely preserved. I will never forget the way both teacher and students fervently adhered to the tiniest detail of preparation and procedure. A spirited woman in her eighties, Nakamura Sensei repeatedly reminded her students that the practice of tea was akin to *zazen*, or seated meditation. Tea required the same focus, the same meticulous attention. The source of this focus was mindfulness with respect for the natural as well as manmade world. Students were taught to walk in the tea room as if their first step were their last.

Figure 2

Nakamura Sensei

After the lesson, Nakamura Sensei made me a bowl of the powdered green tea. It was unlike anything I had ever tasted. Its fragrance was fresh and grassy, and the slightly bitter flavor was balanced by the Japanese sweet she served me. I felt renewed, alert, and, after a second bowl, totally captivated, as much by the ceremony as by the beverage. The movements of host and guest seemed so naturally choreographed. The gentle yet precise motions of purifying the tea scoop, ladling the water, and whisking and serving the tea prompted me to ask Nakamura Sensei if I might come again and take lessons.

But how frustrated I was by those first lessons! It took months to assimilate etiquette and procedures, and even longer to master simple technique. With no cushion between my feet and the tatami mat, the hours I spent sitting on my heels seemed at times endlessly excruciating. I thought of quitting, but after years of Zen meditation I was familiar with the initial difficulties of mastering a discipline, so I stayed.

As I became more used to the posture and began to understand the poetic logic of the tea ceremony, I realized that the more aware I was of every movement, the more my actions seemed to flow, and the more I felt a sense of unity—as much with the utensils as with the guests and the soft mood of the room. It was as if I were meditating, though no meditation was involved. I learned then that tea could be a means of achieving a sense of keen satisfaction. With the mind in a clear and harmonious state, peace could enter the room and my heart. There were mistakes, of course, and occasional confusion, but these were usually minimized by the calming sound of the steaming kettle and the taste of an elegant bowl of tea. At such times, the stoic, encouraging figure of Nakamura Sensei allowed me to maintain my confidence in practice. Although she retired seven years later and went back to Japan, I still felt her presence whenever I returned to Green Gulch.

Like meditation, an appreciation of tea and the practice of ceremony restore a vital sense of awareness, grace, and beauty, qualities so often missing in our age of technology. The art of ceremony offers a way to bring deliberate attention to our actions, allowing us to create a "removal" into rare tranquility. As D. T. Suzuki has said, it "wonderfully lifts the mind above the perplexities of life." [1] Such a removal diminishes the everyday stresses of living that harass the body and mind and add to the difficulties that so often weave their way into our relationships with one another.

By ushering us into the present, the procedure of a tea ceremony offers an experience of the moment-to-moment flow of life and our relationship with all things. Cultivating the earth and harvesting what we plant helps us see that flowers and the hands that arrange them are one, and recognize that we are as responsible for our own well-being as for the world in which we live. Although we may often feel as helpless about personal anxiety as we do about the depleted ozone or the plight of the homeless, we can develop habits that encourage an inner quiet and a sense of wholeness. Thus "removal" is not an escape, but a way to gain greater perspective on our existence. Practice of the tea ceremony provides us with a better way to understand action and reaction. There are no errors in making tea if the attitude of mind is that of true hospitality. Host and guest will naturally know the harmonious thing to do. This kind of insight comes from commitment, from calm and focus. The ceremony brings about a certain mood, and, through the cooperation of host and guest, the deeper meaning is revealed. In a serene atmosphere, as we let go of the ceaseless weave of our thoughts by just sitting and allowing the breath to settle, we realize the connectedness of everything about us. Through a setting of quiet simplicity, tranquility becomes part of the ceremony itself.

I soon learned that to practice tea required not only a respect for its mystique but also a familiarity with its history. In tracing the development of the tea ritual and its significance from East to West, I discovered a rich and fascinating evolution.

---

[1] Suzuki, *Essays in Zen Buddhism*, page 366

Most historians agree that the cultivation of tea began in China at least four thousand years ago, and that it was first used both medicinally and as a condiment. However, shrouded in folklore and academic speculation as the subject is, no one knows when tea was first used as a beverage. Tea became a way of celebrating life through the vision of the eighth century Chinese practitioner Lu Yu. He improved the method of drinking tea by showing how it should be made, and how—through the careful selection and preparation of place, utensils, water, and tea—a simple ceremony could be created in which beauty and harmony were realized.

In China, Zen monks later made a ritual out of drinking tea in front of the image of their first patriarch, Bodhidharma. Japanese Buddhists, who had traveled to China throughout the Tang dynasty (A.D. 618–907) in search of enlightenment and cultural exchange, brought tea to Japan, and by the end of the twelfth century tea was widely appreciated there. [2] In the sixteenth century, Sen Rikyu developed *Chado*, the Way of Tea, and brought to it the spirit of art, hospitality, morality, and philosophy.

First enjoyed by monks in monasteries and aristocrats at court, tea took on cultural significance with spiritual as well as political ramifications. In ancient China, Taoists believed tea to be an elixir; in Japan, it became an intrinsic part of the culture. In the days of Rikyu, whose patron was the powerful shogun, Hideyoshi, strict class distinctions were observed, but those who wished to enter a tea room had to divest themselves of all armaments and rank. To encourage humility among participants in the ceremony, the entrances to a Japanese tea house were reduced in height, so that host and guests had to lower their heads to enter. The entryway for guests was so small that they had to crawl in, which meant that the armed samurai were obliged to leave their swords outside.

The use of tea spread from East to West in the early sixteenth century. The desire for tea in the West soon gave rise to elaborate afternoon rituals, which reached their zenith in Victorian times and are still practiced today—

[2] Varley and Isao, *Tea in Japan*, page 234

albeit with less formality. The popularity of these rituals, however, often obscured the health benefits of the beverage itself.

In researching the various methods of processing tea and herbal beverages, as well as the numerous varieties and their availability, I looked into ancient and current claims about the health benefits of tea. Now that the connection between mind and body and between health and healing has been more widely recognized in the West, there is a much greater appreciation of the curative properties of tea and its herbal counterparts. Recently, Westerners have begun to profit from the Chinese art of healing with tea and herbs. From claims of medicinal value in ancient China to present-day research, tea has remained unsurpassed as a soothing, refreshing, and healing beverage.

Today, more tea is consumed worldwide than any other beverage except water. Since 1960 the sale of tea has doubled in the United States, a phenomenon that reflects as much fascination with the ritual and ceremony of tea as with its consumption.

Ritual is part of individual human life as well as that of society. In Francis Ross Carpenter's introduction to Lu Yu's *The Classic of Tea*, ritual was to the Chinese never an end in itself, but "the behavioral expression of an inward ethic" that intensified one's "belief in the ethic." Deference to ritual paved the way for genuine respect: "Ritualistic acts of graciousness or politeness showed the path to peace and harmony and love. In many ways, ritual served the same ultimate purposes as law in the West." [3]

Although some rituals have a religious purpose or context, others exist in our everyday actions, both conscious and unconscious. Rituals have been devised for the various passages in the life of the individual as well as the tribe, and even today we observe rituals in eating, mating, politics, and war. Yet modern life in the West has evolved in such a way that we frequently tend to disregard the serious exercise of ritual. Its importance has diminished. But we can reclaim ritual in our personal lives and recognize how, by providing a meditative experience, the ritual of a tea ceremony can help

[3] Lu Yu, page 6

organize our scattered thinking and draw us into the present. With this aspiration in mind, I describe practical ways to create tea rituals in Chapter V, "Ways to Tranquility."

I was drawn to the Japanese ceremony because of its refinement, beauty, and the unique removal it provides from the concerns of a busy life. Throughout the world, people from all walks of life have studied the Japanese Way of Tea, or *Chado*, from qualified Eastern and Western teachers. For most Westerners, however, traditional Japanese tea practice may present two problems.

The first is the linguistic and cultural barrier. Japanese is a difficult language to master, and the culture is very different from Western ones. Nevertheless, it is not hard to learn the required phrases and understand the basic instructions. And as one walks along the stone path of a hand-constructed tea house on a quiet afternoon or observes a reflection on the lid of a red-lacquered tea container, one need not be part of the culture to be moved by these simple pleasures.

The second problem is physical: most people have difficulty kneeling and sitting on their heels for long periods while practicing the tea ceremony. Fortunately, a table-style tea has been initiated in Japan, so that now a Way of Tea can be faithfully realized by anyone regardless of physical limitations. Although sitting at a table would never replace the naturalness and intimacy of the tatami mat of a Japanese tea room, table-style tea offers older students and those who have difficulty kneeling a way to appreciate the traditional ceremony. This adaptation allows even those in wheelchairs the opportunity to study tea.

Today in Japan, Dr. Sen Genshitsu, the fifteenth descendant of Sen Rikyu and former Grand Master of the Urasenke tradition of *Chado*, tells us that "a world of peace can start with just two individuals." His modern approach, which encourages innovation while preserving tradition, is also illustrated by the new spirit of *Sabie Zen*, a movement organized by the Sabie Cultural Institute in Kyoto (see the glossary). Its aim is to create a new understanding of the beauty in *Chado*, and it includes the use of modern works of art from the West as well as the East.

In considering this approach, I came to see how a set of simple rituals could be derived from the traditional principles of harmony, respect, purity, and tranquility, yet be more suited to Western lifestyle and habits. In Part Two, Practical Adaptations, I offer a version of a Japanese table-style tea, as well as adaptations of Chinese and English ceremonies that anyone may do on one's own.

Adaptations can lead to a genuine understanding of the practice of tea ceremony. A meaningful tea ceremony may be conducted not only through time-honored procedures within a traditional tea house, but in any number of settings and under a variety of circumstances. Whenever we can pause for a few minutes, alone or with like-minded others, it is possible to create a removal from everyday concerns. The key is to provide the time, develop patience, and remember that the magic of tea improves not only our lives but the world in which we live. As Lu Yu said centuries ago, ritualistic deference "paves the way for genuine respect" and helps us find the "path to peace and harmony and love." [4]

[4] Lu Yu, page 6

An Overview of Tea Cultures

The desire to know more about Chado and the aesthetic that lies behind its ceremony more often than not spawns curiosity about the history of the beautifully crafted utensils used in tea practice. Keeping track of who was responsible for fashioning a certain style of tea caddy, or what generation of tea master might have carved a particular tea scoop may prove frustrating to a student without a proper source of reference or fluency in the language. Such particulars though, are but small parts of a great, global mosaic of tea cultures, one that demands and deserves attention. The frequent use of Rikyu's name in connection with "the spirit of tea" in Japan, naturally leads to reading about the history of tea, its discovery in China, and the genius of Lu Yu. While doing so, one is easily enmeshed in an enormous weave of legend and myth, the complexity of which vividly contrasts with ordinary, sensible things such as Lu Yu's appreciation of the quality of the water used for tea and Rikyu's disdain for ornate utensils. Yet, a familiarity with tea cultures enables serious students to glean insights that continually enrich their practice.

In most of the world, tea is connected in people's minds with China because the tea plant, *Camellia sinensis*, was first cultivated there. This evergreen shrub is indigenous to parts of China and India. In its wild state, the bushes grow higher than thirty feet and produce fragrant, cream-colored blossoms; for cultivation they are pruned to somewhere between three and five feet in height. All tea—excluding herbal beverages—is produced from Camellia sinensis.

The cultivation of this wonderful gift of nature has spread widely over the centuries. The beneficial and pleasing qualities of tea lend an otherness to the beverage, offering a kind of spiritual renewal that has inspired a number of highly refined ceremonies in cultures throughout the world.

Figure 3

Camellia sinensis

In China

One of the legends about tea claims that it was first brewed in China around 2700 B.C. by Shen Nong, the Divine Cultivator, who was one of five mythical sovereigns of the time. Shen Nong was also revered as the inventor of agriculture and medicine. It is said that once when he was outdoors boiling water (which he wisely believed prevented illness), a few leaves from a nearby tea plant fell into the steaming pot. When he sampled the results of this accident, he was delighted with the taste and became convinced the brew had medicinal value as well. He began to use tea as an antidote to the poisoning that sometimes occurred in his experimentation with other plants.

Not every legend is so down-to-earth. In Buddhist lore, the tea plant is believed to have originated when the early Zen patriarch Bodhidharma came to China from India in the sixth century. It is said that he meditated for nine years before a wall. To keep from falling asleep, he cut off his eyelids and threw them to the ground; where they landed, a tea plant sprang up. However fanciful this particular legend may be, tea was, and still is, used as a stimulant to aid monks during their long hours of meditation.

The tales about tea are many. According to one account, the cultivation of wild tea plants began in the Szechwan province of China in the third century and spread down the valley of the Yangtze River. In another story, tea was discovered as early as 1200 B.C., when tribal heads of Szechwan presented it as a tribute to King Wen, the author of sophisticated essays on the *I Ching*.

Early Chinese records indicate that tea was rare; it was prized by emperors and used as a reward in court. Methods of preparation were still crude, however. Freshly picked green leaves were simply dried, crushed, and then boiled in water. In later times, after about 300 B.C., they were steamed, crushed, pounded, and compressed into cakes. A bit of the cake might then be broken off and boiled with other ingredients, such as rice, salt, ginger, spices, milk, orange peel, and even onion. By the time of the Tang dynasty (618–907), the preparation of tea had evolved into an art. The consumption of tea in China had become widespread, and tea rituals had been improved thanks to the wisdom of Lu Yu.

Abandoned as a child, Lu Yu was adopted by a Buddhist priest. In his youth, while brewing tea for his foster father, he came to see that harmony could be experienced in simple yet precise service. Though trained for the priesthood, he eventually rebelled and, seeking independence, began to wander. In *The Romance of Tea*, William H. Ukers tells how Lu Yu earned his living as a clown, something he had longed to do. But no matter how much he delighted crowds, he was unhappy. He yearned for knowledge and understanding. One admiring official recognized his desire to learn and offered him the use of a vast library of ancient wisdom. This gave Lu Yu the opportunity to create something meaningful, which was realized when tea merchants approached him for a book on tea. They wished to compile all that was known about growing and producing tea in order to raise standards and heighten the enjoyment of the beverage. Influenced by Taoism, Lu Yu saw how tea ritual could express the same harmony and order that could be found in nature. With this in mind, he wrote *The Classic of Tea* and became the nation's acclaimed apostle of tea.

In his book—which took twenty years to complete and was published in A.D. 800, four years before his death—Lu Yu extolled the benefits of tea and elevated its consumption to a new, aesthetic height. With a keen awareness of nature, he sketched an idealized landscape of clouds, mists, and mountains, yet offered a practical appreciation of tea. The book became a tremendous success, recognized by both merchants and the literati.

Lu Yu teaches us how to manufacture and use tea, arrange the utensils, and brew tea correctly. His directions are strict, yet poetic. For instance, in the first of the three stages of boiling the tea water, the water is to be "like fishes' eyes" and "give off but a hint of a sound. When at the edges it chatters like a bubbling spring and looks like pearls innumerable strung together, it has reached the second stage. When it leaps like breakers majestic and resounds like a swelling wave, it is at its peak." [1] The Japanese scholar Okakura Kakuzo recalls Lu Yu's description of how the tea cake is "roasted before the fire" until it becomes "soft like a baby's arm" and is then "shredded into powder between pieces of fine paper." The salt is put into the water during the first boil and the tea during the next, and a dipper of cold water is poured into the kettle as it comes to the third boil. This is to "settle the tea" and revive the "youth of the water." [2]

Later, Lu Yu wrote about where the best water could be found—"from midstream on the Yangtze at Nanling"—and warned that "near the bank" the water was often brackish. His discrimination was unerring. Once, on a boat trip, he was served water that supposedly came from that prized river. He sipped it but complained that it had been taken from near the bank. A servant swore that was not true, so Lu Yu drank again and agreed that it could have been taken from midstream, but said it was nevertheless mixed with other water. The servant finally admitted that his boat had rocked and spilled some of the water from the jar, and that he had refilled it closer to the bank.

Lu Yu was as particular about the preparation of tea as he was about the selection of water. In *The Chinese Art of Tea*, John Blofeld tells the story of how Lu Yu's adoptive father was so fond of the tea prepared by his son that he gave up tea when Lu Yu left home. When his abstinence came to the attention of the emperor, the ruler summoned the father to the capital and offered him an excellent cup of tea brewed by a lady of the court. Not knowing who had made the tea, the old man sipped and then respectfully set the cup down. It was not as good as Lu Yu's. The emperor felt the old man was merely showing off. To expose him, he ordered Lu Yu—who, unknown to his father, was at court—to brew tea for an "unnamed guest." When the old man sipped this cup of tea, he delightedly exclaimed that it was so superb that even his son could not have done better. Father and son were then reunited. [3]

[1] Lu Yu, page 107
[2] Okakura, page 26
[3] Blofeld, page 7

During the Song dynasty (960–1280), tea tasting grew to be a popular pastime among many classes of people, much to the disdain of aesthetes. Aficionados would gather to display their expertise in detecting the region where the teas were grown and the origin of the water. As the enjoyment of tea became more widespread, demand for it grew until it was available not only to aristocrats and wealthy merchants, but to peasants as well. Once considered a luxury, tea was now a necessity. Tea houses were popular in Hangchou, the great southern capital. Locals could enjoy not only tea there, but also soups and other foods, while admiring flower arrangements or art displays. Among the affluent, the consumption of tea developed into a ritual, with some ceremonies demanding great learning and skill. These rituals eventually spawned private tea houses, which were built in the gracious gardens of aristocrats and government officials.

By the time of the Song dynasty, the method of processing and preparation had changed. Dried tea leaves were ground into a fine powder to which hot water was added, and a delicately designed split-bamboo whisk was employed to blend the tea. Salt was no longer used. Different and more elaborate ceremonies were developed using powdered tea, and the need arose for newer and more elegant utensils to accommodate these changed methods. Celadon and *temmoku* ware (see the glossary) produced during the Song dynasty are still considered among the finest ceramics in China's artistic history.

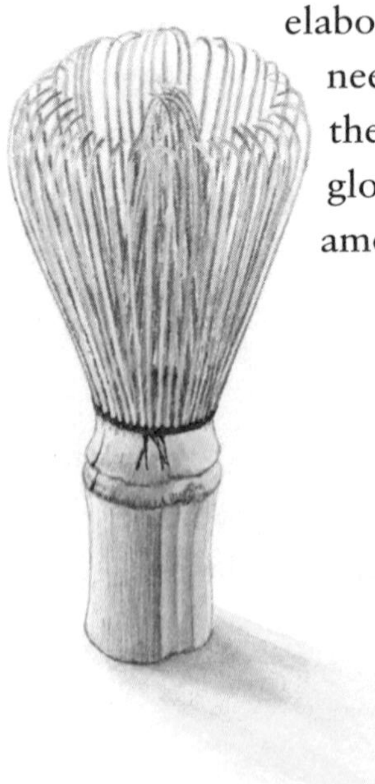

Figure 4

The whisk

From the early Tang to the Song dynasty, the evolution of religious beliefs dramatically affected the tea ritual. Along with Confucianism, Taoism, and Zen Buddhism, the poetic use of tea became a path to self-realization. According to Okakura, the Songs in their "tea-ideal" differed from the Tangs even as their notion of life differed, in that they sought to actualize what their predecessors tried to symbolize. The Taoist conception that immortality lay in eternal change permeated all their modes of thought.

> *It was the process, not the deed, which was interesting. . . . the completing, not the completion, which was really vital. . . . the Buddhists, the southern Zen sect, which incorporated so much of Taoist doctrines, formulated an elaborate ritual of tea. The monks gathered before the image of Bodhi Dharma and drank tea out of a single bowl with the profound formality of a holy sacrament. It was this Zen ritual which finally developed into the tea-ceremony of Japan in the fifteenth century.* [4]

Taoism, Zen, and the Way of Tea

In the broadest sense, Taoism, from which many Zen tenets were drawn, is a reflection of the natural workings of the universe. According to the Taoist view, the universe is characterized by *yang* and *yin*—male and female, positive and negative principles that work in balance to maintain a certain order that is evident in creation and destruction. This view sees the opposites as interdependent rather than conflicting. Living in accord with the two principles is the way of the Taoist. Night turns to day; water, even

[4] Okakura, page 29

at a trickle, wears away solid rock. By understanding the constancy of change, a Taoist follows the path of least resistance and endures both good and bad circumstances by remaining unattached to desires and to antagonism that generates desires. Embracing the Tao requires training in meditation and, in a few sects, the martial arts. Taoists believe that discipline, devotion, and a poetic sensibility pave the way to enlightenment, understanding, and freedom from suffering.

Taoism has numerous rituals for transforming a person in accordance with its views. Some of these rituals, with roots going back to the beginning of Chinese society, show clear shamanistic influences. One such ancient ritual preserved in Taoism is a dance called the Steps of Yu, [5] essentially a way for human beings to integrate themselves with the cosmos. It was one of the rituals in which the performance of prescribed movements was believed to lead to higher states of being. Serving a similar purpose would be the chanting of *sutras* and the performance of certain rites during festivals.

At the same time in China, ritual took other forms. The Confucianist believed that by performing rituals and sacrifices, the emperor could maintain his mandate to rule and, in turn, keep his people in harmony with nature. The cultural attitude that admires an emperor before an altar holding up a vessel of wine to heaven would also admire a refined person holding up a vessel of tea before a vase of flowers and a scroll with words of wisdom. Looking at the tea ritual and taking into account Confucianism and old shamanistic practices, we can see how Taoism influenced Zen and tea through the appreciation of nature, the insistence on direct individual experience, and the significance of the moment.

Understanding the link between Taoism and Zen, I came to see the effect of Zen on tea practice. Zen offers a way to see into the heart of reality. It produces an enlightened state, a communion with the universe that reveals a true tie to the present, to the now, rather than to the pain of the past and the fear of the future.

[5] Wong, Eva, *The Shambala Guide to Taoism*, pages 11–13

The word Zen comes from the Sanskrit word *dhyana*, meaning "meditation." Although it involves meditation, it is not in itself meditation. It is a way to expose the fundamental essence of the true self beyond intellectual conceptions. For a Western understanding of Zen, one can turn to Alan Watts, who wrote and lectured extensively on the subject.

In comparison with many other religions, which have scriptures containing volumes of subtle philosophical discourse, Watts finds Zen straightforward and practical. He maintains that the spiritual answers we seek are usually difficult to find because we ponder too much and are constantly looking in shadowy corners for what is directly before us. Lecturing on Japanese haiku, Watts told of a Zen master who warned that if we want to see into reality, we have to "see into it directly." When we think too much, "it is altogether missing."

Zen answers profound questions with simple, everyday truths. The difficulty in talking about Zen is that every attempt to explain it tends to make it more obscure. According to Watts, one of the best ways to understand Zen is to recall a certain moment in life when you were aware of being unusually alive. That aliveness was intense, but only to the extent that it was not grasped. If you allow this aliveness to simply happen—a matter of one or two seconds, usually—you are in the moment. You feel what you feel, and then let it go. For Watts, this letting go is

> the whole art. That's why our eyes see best when they brush across things and do not stare in a fixed gaze . . . the reason is that whatever is momentous, living and moving, is momentary. Minute by minute our experience moves along without return. . . . We are at accord with this experience to the degree we move with it, as the mind follows music, as the leaf goes with the stream. Yet this is to say too much, for the

moment we stop to philosophize about it, to make a system of it, we have missed it. [6]

Enlightenment was pursued by the most direct means possible in Zen Buddhism. The actual practice of meditation superseded any adherence to doctrine or dogma. Meditation, it was believed, would facilitate the experience of seeing into one's true, or original, nature. As an adjunct to this practice, tea ceremony became a way of Zen because it required a focus on the present, on the awareness of body and mind, much like meditation. Tea practice encouraged harmony with nature and the experience of the natural self.

The halcyon days of the Chinese tea ritual during the Song dynasty came to an end with the Mongol conquest. China fell into disarray, and it was not until the Ming dynasty (1368–1644), after the Mongols were driven out, that the practice of tea regained its significance. Soon after, tea began to be exported to foreign nations. However, by the seventeenth century, whisked tea had been all but forgotten in China. Okakura mentions how a Ming commentator was at a loss even to "recall the shape of the tea whisk mentioned in one of the Song classics." [7] Instead, loose tea leaves were steeped in a cup or bowl of hot water, much as they are today. It was during the Ming dynasty that the Yixing teapot was created. Yixing was long known as the pottery capital of China. Prized in Europe, the simple elegance of the small, reddish earthenware containers became a model for Western teapots.

In modern times, the fall of the Qing dynasty in 1911 brought civil war to China, and the Communists' Cultural Revolution attacked old traditions. Certain aspects of tea ritual, such as the *gong fu* ceremony, continue in such places as Hong Kong, Taiwan, and Singapore, and tea survives in China as a delicious and nurturing national beverage. For no matter how it is served, a bowl or cup of tea is a symbol of warm hospitality, offered with respect in even the poorest of households.

[6] Watts, Alan, *Haiku*
[7] Okakura, page 30

In Japan

Tea is thought to have been introduced in Japan around A.D. 729, when the Emperor Shomu invited a hundred monks to have tea at his palace. The tea is believed to have been brought from the Tang court in China by Shomu's ambassadors, [8] and before long tea drinking was adapted by the Japanese imperial family. Some Japanese Buddhist monks who had gone to China to study returned with tea seeds. One of these, Eichu, had spent nearly thirty years in the Chinese capital of Ch'ang-an. [9] The seeds he brought in 805 were planted at monasteries in Omi and produced a tea that is said to have been served in 815 to the visiting emperor, Saga. Saga, who greatly admired Chinese art and poetry and had a taste for all things Chinese, made the beverage popular at court events and poetry gatherings.

As Buddhism began to spread across Japan in the late twelfth century, Eisai (1141–1215), a middle-aged Japanese priest who had been studying in China, returned with both *matcha*—powdered green tea—and a greater awareness of the practice of Zen. He is believed to have brought tea seeds that were planted in the Hizen district and then transplanted at Hakata in Kyushu, where his first Rinzai Zen temple was built. Tea was also planted at Kozanji Temple, northwest of Kyoto. Much later, tea from Kozanji was grown in the Uji district near Kyoto, and in time this tea came to be considered the finest in the land.

Eisai, a keen observer, was one of the first in Japan to write about the health benefits of tea—how it helps stabilize blood pressure, cures boils and beriberi, and contributes to a general peace of mind. His observations were presented to the shogun Minamoto Sanetomo, along with some powdered green tea. The ruler was encouraged to drink tea not only because it was a fine beverage, but also because it was an excellent remedy for a hangover. Zen monks also took advantage of its benefits and drank tea for stamina

[8] Tanaka, page 23
[9] Varley and Isao, page 6

during long hours of meditation. By the late Muromachi period (1336–1568), the Way of Tea had become a means of encouraging Zen in society. The self-discipline required in the practice of tea attracted the ruling samurai to such a degree that the tea ritual came to complement the militarism of the era.

The Japanese were familiar with the Chinese tea ritual. They followed the development of the tea ceremony elevated by Lu Yu during the Tang dynasty and later adapted the style of the whisked tea of the Song dynasty. Because they succeeded in warding off the Mongols in the late twelfth century, the Japanese were able to preserve and cultivate their traditions. One of these was the ritual of powdered tea, which was finely ground from unfermented leaves that had been dried and stored for at least six months.

Like the Chinese, the Japanese held *tocha*, or tea–tasting contests, in which regional teas were distinguished and judged. But these contests were eventually ruined by gambling: large bets were placed on who could best identify the different teas and the regions in which they were grown. A. L. Sadler writes that lords and wealthy merchants laid great stakes of "fine incense, gold dust, valuable silks and brocades, and armor and swords, though when they won these things they thought it beneath them to keep them, but gave them as presents to the Dengaku players and dancing girls." [10] In *The Tea Ceremony*, Sen'o Tanaka mentions a famed account of a *tocha* gathering where a host "offered one hundred different kinds of stakes, while another piled up one hundred rolls of dyed silk before the players, and a third provided ten different kimono as prizes. In this way the hosts could flaunt their own wealth as well as decorate their homes with precious works of art on these occasions." [11] *Tocha* became so extravagant that it was banned in the fourteenth century by Ashikaga Takauji, the first Muromachi shogun.

By the end of the sixteenth century, the Japanese had refined the tea ritual which had come to embrace so many of their arts. They fashioned

[10] Sadler, page 4
[11] Tanaka, pages 26–27

unique utensils, including iron kettles, tea bowls, water jars, incense burners, tea caddies, and tea scoops, and produced exquisite paintings and calligraphy that were proudly displayed in their tea rooms. Utensils and scrolls from China were also highly esteemed and were treated with great deference.

Figure 5

Song bowl: Temmoku type

As a spiritual practice, the tea ritual reached its peak in Japan during the sixteenth century. A Zen priest named Murata Shuko (1422–1502) is credited with transforming what had become an extravagantly performed ceremony into an unpretentious and tranquil one. He brought a singular mindfulness to the performance of the tea ceremony, believing that enlightenment could occur in acts of simple, everyday life. Shuko favored using humbler, less refined tea utensils from his native country together with the highly prized ones imported from China. He felt beauty in the obscure—preferring, for instance, to see a moon that was partially shrouded in clouds rather than a full one in a clear sky.

During the rule of Shogun Ashikaga Yoshimasa (1436–1490), the Japanese tea ceremony, based on the practices of tea drinking by Zen monks, the aristocracy, and the samurai elite, took a definitive turn. Yoshimasa found the aesthetics of tea a welcome diversion from war and domestic strife, and at his villa in Kyoto (which later became the famous Silver Pavilion), he devoted a small room to the tea ceremony. The dimensions of this room, nine or ten feet square, are still seen as ideal. The shogun was impressed with Shuko and took lessons from him. It is said that Shuko devised a special tea ceremony at the shogun's request and consequently became one of the first tea masters of Japan.

According to Okakura, by 1600 tea had become a religion of the art of life. "Teaism" was founded on the appreciation of the beautiful in the ordinary elements of life: Tea "inculcates purity and harmony, the mystery of mutual charity, the romanticism of the social order. It is essentially a worship of the Imperfect, as it is a tender attempt to accomplish something possible in this impossible thing we know as life." [12]

To Okakura, the philosophy of tea was "not mere aestheticism" in the ordinary sense of the word. Combined with ethics and religion, this philosophy expressed the whole Japanese point of view on humanity and nature. Moreover, tea was hygienic in that it enforced cleanliness; and it was economical in that it led to a preference for comfort in that which was simple rather than complex or costly. Tea provided a "moral geometry":

> *The tea-room was an oasis in the dreary waste of existence where weary travelers could meet to drink from the common spring of art–appreciation. The ceremony was an improvised drama whose plot was woven about the tea, the flowers, and the paintings. Not a color to disturb the tone of the room, not a sound to mar the rhythm of things, not a gesture to obtrude on the harmony, not a word to break the unity of the surroundings, all movements to be performed simply and naturally—such were the aims of the tea*

[12] Okakura, pages 3–4

ceremony. And strangely enough it was often suc-cessful. A subtle philosophy lay behind it all. Teaism was Taoism in disguise. [13]

The tea ritual devised by Murata Shuko in the fifteenth century became the foundation of *Chado*, the Way of Tea. Shuko's work inspired a famous tea master, Takeno Jo-o (1502–55)—a Zen disciple who had studied with two of Shuko's students—to conceive of *chanoyu*. Although cultured, Jo-o shunned any display of wealth and found the austerity of a grass hut to be the best setting for tea. *Chanoyu*, a humble yet innovative ceremony, required total attentiveness. The word literally means "hot water for tea," a deceptively simple translation; actually, *chanoyu* embodies all that tea means to the Japanese. Sadler describes it as a "household sacrament of esthetics, economics and etiquette." It involves both the practice of tea and a way of life that integrates attitudes of modesty and awareness.

In sixteenth-century Japan, tea meant different things to different people. Before their adoption of chanoyu, drinking tea was already an amusement for some of the nobility. Connoisseurs had made tasting competitions popular as in *tocha*. Merchants had used *chanoyu* as a device to awaken spirituality ever since Shuko laid the foundation for the tea ceremony. Business and politics were sometimes deliberated during the taking of tea, and in other instances the bartering of treasured scrolls or utensils took place. All of this was hardly true to the Way of Tea. It took the teachings of Sen Rikyu (1522–91), a student of Jo-o, to restore the ceremony to its ageless principles.

Just as the work of Lu Yu dramatically increased the popularity of tea in China, so Rikyu's teachings affected the Way of Tea in Japan. It was not until Shuko, Jo-o, and especially Rikyu established *wabi* as the aesthetic ideal that the tea ceremony was further refined into a way of Zen.

Wabi, a word of many subtleties, also seems to escape translation. It is an aesthetic that embodies the beauty of the imperfect. *Wabi* "means to

[13] Ibid., pages 33–34

transform material insufficiency so that one discovers in it a world of spiritual freedom unbounded by material things." [14] Used to describe an attitude, an atmosphere, a person, or some material object, the term denotes a certain humility, solitude, unpretentiousness, and austerity. It is a preference for the bud instead of a full bloom. A story is told that the young Rikyu once cut off the small handle of a water container to make it more *wabi*. His tea master, Jo-o, felt that the taste of *wabi* could be savored in the following verse by the poet Fujiwara no Teika:

> *Looking about*
> *Neither flowers*
> *Nor scarlet leaves,*
> *A bayside reed hovel*
> *In the autumn dusk.* [15]

To Jo-o, the spell of a fading autumn day by an empty shore was a kind of spartan beauty, without vivid color or light. Fujiwara no Teika weaves another spell in a poem treasured by Rikyu as an ideal image of *wabi*:

> *To those who wait*
> *Only for flowers*
> *Show them a spring*
> *Of grass amid the snow*
> *In a mountain village.* [16]

Rikyu had different names throughout his life; he was born in 1522 as Yoshiro. Growing up in a merchant family in Sakai, a thriving port near Osaka, he studied tea and became an outstanding pupil of Takeno Jo-o. At

14 Varley and Isao, page 96
15 Ibid., page 199
16 Ibid., page 200

eighteen he was given the name of Soeki by his Zen master. As his reputation grew, he attracted the attention of Nobunaga, the military ruler of Japan, who became his patron. When Nobunaga died in 1568, he was succeeded by Toyotomi Hideyoshi, a genius in war and politics. Rikyu, already a widower and a father, gained an even greater patron in Hideyoshi.

Through him, Rikyu met the emperor, who in 1585 gave him the title of *koji*, or "enlightened recluse." With this honor, he took the name of Rikyu and gained even greater influence with Hideyoshi. Unfortunately, this proved to be his undoing, for within Rikyu there existed an odd dichotomy. On the one hand, he was now remarried and a man at ease in the world of wealth and aristocracy. On the other, he was a spiritual seeker who not only imparted *wabi* sensitivity to tea, but also popularized the idea of a smaller tea room and utensils that were "nothing special." Rikyu felt that the smaller the tea room, the better it could create intimacy between host and guest. Nothing could enter the tea room except the bare self. With hearts and minds wide open, unity could be experienced.

As with Jo-o, Rikyu's ideal tea room was a separate hut set in rustic surroundings and built of materials suggesting simplicity and humility. Paradoxically, these small, apparently artless structures required great skill and a sense of craft, and in time, they became extremely costly to build.

Hideyoshi, who had come from humble circumstances, was attracted to Rikyu's Way of Tea and was constantly amazed at his subtle skill and his approach to elegance and beauty. He often tested that skill; a story is told that he once challenged Rikyu by presenting him with a long branch of plum blossoms. With only a golden bowl and a bucket of water, he was expected to display the flowers in the alcove of the tea room. Rikyu poured water into the bowl, shook the plum blossoms into it, and displayed them floating in the bowl.

However much Hideyoshi might have been drawn to such aesthetics, he was still inclined to use the tea room for political ends. His desire to use gold utensils and cover a tea room's walls in gold leaf seemed far from *wabi*.

Although Hideyoshi's Golden Tea Room proved impressive and a work of art that even Rikyu admired, Hideyoshi's flamboyant style did conflict with Rikyu's *wabi* approach. It is believed that sometimes, bothered by the attitude of those who used tea ceremony for business or as a means of showing off wealth, Rikyu would recite a favorite poem by Ji Chin, a Buddhist priest:

> *What a pity it is*
> *That the Pure and Perfect Law*
> *We should keep unstained,*
> *Is by men so frequently*
> *Made a source of worldly gain.* [17]

Perhaps the poem offended Hideyoshi, who had done much to promote the status of the tea master. If so, it must have added to the tension between the two men, a tension that increased when Rikyu felt that Hideyoshi's plan to invade Korea after his completion of national unification was unwise.

About this time, Hideyoshi was attracted to one of Rikyu's daughters, a beautiful widow. But she rejected him, and Hideyoshi came to believe that the rejection had been influenced by her father. The trouble reached a climax when a life-size statue of Rikyu was erected on top of one of the gates of the Daitoku-ji temple in Kyoto. Hideyoshi thought this was outrageous, for it meant that he, or anyone entering the temple, would have to pass beneath the statue's feet. How much Rikyu was responsible for the statue is not known. But such elevation of him was taken as a direct affront, and the rumors and intrigue that followed led Hideyoshi to demand the death of Rikyu. So it was that in 1591, Rikyu, at seventy years of age, was placed under house arrest at his Jurakudai residence in Kyoto. One story tells of his going into his tea room for a last time, arranging a flower, quietly preparing

[17] Tanaka, page 42

a bowl of tea, then drawing his sword and committing *sepukku* (ritual suicide). In another version, he was summoned to his residence in Kyoto, where he killed himself. Today in Kyoto, at the Jukoin subtemple within the Daitoku-ji complex, his grave may be found, dignified by a tall stone lantern that rises above all who approach it.

More than four hundred years later, Rikyu is still venerated. He brought a quiet elegance to the tea ceremony and instilled in it the principles of harmony, respect, purity, and tranquility. These, together with his basic guidelines, continually inspire a calm attitude of mind in tea practitioners and provide a foundation for serious practice. The essence of Rikyu's *chanoyu* is reflected in a poem of his:

> *Tea is naught but this:*
> *First you heat the water,*
> *Then you make the tea.*
> *Then you drink it properly.*
> *That is all you need to know.* [18]

In India

Outside of China and Japan, most of the world knew little about the cultivation of tea until the eighteenth century. China had managed to guard its monopoly closely for nearly ten centuries, having made it a crime punishable by death to reveal any secret regarding the cultivation or processing of tea. India was among the countries that began to import seeds and plants from China in the late eighteenth century, although it was initially unsuccessful in trying to grow tea plants. It was suspected that seeds exported from China might have been boiled to prevent their germination, and often both seeds and plants were so badly packed as to render them useless on arrival.

[18] Sen Soshitsu XV, *Tea Life Tea Mind*, page 79

The monopoly was finally broken when a Scottish botanist named Robert Fortune, an agent of the East India Company, visited China in 1848. It was not Fortune's first trip. On this occasion, however, two Chinese companions helped disguise him as a Chinese merchant. Incredible as it seems, he was able to observe farming and harvesting methods, successfully smuggle out seeds and plants, and return with instructions on growing and processing tea. Ironically, India had always had its own native tea plants, but only the aboriginal Indians were aware of them. Once these indigenous plants were finally recognized and successfully cultivated, India went on to succeed China as the world's greatest exporter of tea.

Many writers have told the tale of how Indian tea was first "discovered." Apparently, in 1820 another Scotsman, Robert Bruce, found a tea bush growing in a jungle in the Indian state of Assam, north of what is now Bangladesh. His brother Charles, who lived in Assam, grew some of the plants in his own garden in Sadiya, near the Burmese border. Some of the leaves and seeds were later sent to a Dr. Wallich, then head of the Botanical Gardens in Calcutta. At first he dismissed the plant as nothing more than a variety of the camellia. Later he became an enthusiastic convert to this Indian tea plant. Since people had been drinking the Chinese variety of tea for centuries however, some growers preferred to keep trying to cultivate it, thinking that the Assam variety might be undependable. From Calcutta's bazaars, Chinese shoemakers and carpenters were brought to Assam to teach local tribal members how to process tea. Unfortunately, the Chinese ended up brawling with the natives and had to be deported. What the growers did not realize was that green and black tea came from the same plant, and that the different colors were due to different processing methods. It was not until Robert Fortune returned with the secrets from China that the truth was uncovered. The productive farming of Indian tea began about the time that the British East India Company lost control of its Chinese trade.

Assam tea became popular in India and was exported to England. At first, English aristocrats complained that it was inferior to Chinese tea. This was probably a matter of preference, although the size of the leaves and how they were processed may have affected their taste. Determined to compete with the Chinese, Indian producers formed the Assam Tea Company and persevered, but it was almost twenty years before they were able to show a profit. Not long after, India became such a successful tea producer that British subjects began to see it as their patriotic duty to drink Indian tea. With this, imports of Chinese tea began to decline.

Calcutta served as the headquarters for tea brokers, shipping agents, packagers, distributors, and plantation owners. By the twentieth century, India had become the largest tea producer in the world. The cultivation and production of tea employed more people there than any other industry. William Ukers observed that the sun never sets on Indian tea: it is either grown or consumed in Europe, the Middle East, Sri Lanka, Asia, Africa, Australia, New Zealand, North America, and South America.

Known as *chai* in India, tea is enjoyed daily by members of every caste and religion. It is also sometimes used as a spice or vegetable. Tea drinking in India never inspired rituals similar to those developed in ancient China or refined in Japan by great tea masters. The idea of drinking tea in a glass half-filled with milk and loaded with sugar, as is customary in India, might have repelled Lu Yu, but it is nonetheless a staple drink and a pleasant custom for hundreds of millions of people. To be sure, there are times when pure tea is enjoyed—usually at tea tastings, where judging various fermented teas for aroma, taste, and quality provides a diversion for the connoisseur. These tastings are conducted with as much attention to the leaves when they are dry as when they have been brewed. The final test lies in color and taste, and these variations become an excuse for passionate and obsessive discussion as well as mere social intercourse. This is a far cry from the aesthetics of Japan, where peace is sought and a communion with nature preferred. As with law and language, however, the Indians have emulated the British in tea ritual to such an extent that in both grand and humble homes, as well as in first-class hotels, the custom of English teatime is observed.

The island of Ceylon, now Sri Lanka, also became a principal tea producer, thanks to a few resourceful planters who refused to be defeated when a terrible blight destroyed their coffee plantations in the 1880s. Tea seeds were brought from India and planted where coffee bushes had stood. Against formidable odds, a tea industry rose out of the ruins of the coffee estates and eventually produced some of the finest tea on the world market.

In Europe

For centuries, there was little or no knowledge in the West of the ancient ways of tea. Along the silk route from Dunhuang in China, tea was carried in the form of bricks to Tashkent and Merv in central Asia. From there, overland caravans traveled into Afghanistan and Persia, then on to Damascus, eventually reaching Europe. In the early seventeenth century, the Dutch East India Company began trading with China and Japan and importing green tea into Europe. Although at first the tea was expensive and sold only in medicine shops, it soon became more affordable and more widely available. Black tea was especially popular in Holland, where the wealthy began to hold tea rituals in exclusive rooms decorated for the purpose. The less affluent formed tea clubs, reserving rooms in public places and sometimes meeting in beer halls. The private tea party became so fashionable that Dutch husbands complained that their wives were ignoring their household duties.

The Dutch East India Company was also successful in developing its own tea plantations in Java and Sumatra with seeds brought from Japan. Later, after years of importing processed tea from China, the Dutch obtained seeds and plants from that country. The figure behind this successful expansion was J.I.L.L. Jacobson, a visionary who learned his trade from his father, a coffee and tea broker in Rotterdam. Although appreciated by the Dutch for his amazing accomplishments, Jacobson had enemies among the Chinese mandarins, who put a price on his head for piratical meddling. It was not until 1878 that the Dutch cultivated Indian tea in Java and Sumatra. By then, Dutch holdings in the Far East were flourishing, even though the Dutch East India Company had become so corrupt that it was later dissolved and made a part of the Dutch colonial government.

Although the French and the Germans were not as fascinated with tea as were the Dutch, tea made a great impression in Russia. Because bringing tea by camel caravan was so expensive, it was at first a commodity enjoyed exclusively by the Russian aristocracy. By the end of the eighteenth century, however, consumption had become widespread, reaching more than three and a half million pounds a year—the equivalent of over six thousand camel loads of tea. This made Russia one of the greatest tea drinking nations in the world after England.

Russians liked brick tea—fermented tea that had been compressed into blocks. Generally, it was of lesser quality, because the leaves were often braced with tea dust. Tea was enjoyed at any hour of the day, its popularity giving rise to Russia's unique contribution to the history of tea: the fabled samovar. After 1830, samovars could be found throughout the Russian empire; eventually, they were seen in affluent drawing rooms worldwide.

The samovar was a graceful and sometimes highly ornate brass, copper, or even silver-plated urn. The sizes varied, but there were two basic kinds of samovar: one in which tea was made, and another in which water was boiled for a matching teapot that was fitted to sit on top.

Fired by a charcoal brazier below, the samovar was a faithful provider. Strong tea was poured into glasses usually encased by metal or silver holders, after which steaming water was added to suit the guest's taste. Lemon or sugar might also be added, though usually not milk or cream. Occasionally, in place of lemon, a spoonful of jam was stirred in. Sugar was sometimes spooned from a bowl, though provincials preferred to hold a cracked piece of sugar cube between their teeth while sipping the strong brew.

Figure 6

**Russian with
a samovar**

As in Russia and Holland, tea took a strong hold in England. When England's trade with China finally began in 1684, tea was exchanged for silver. However, as tea became popular, England, to prevent a serious depletion of its silver reserves, began to export opium to China—without the blessing of the Chinese Emperor. This eventually led to a series of tragic opium wars, the first of which broke out in 1839. At that point, England recognized that it needed other sources of tea. Once the plant was successfully cultivated in India and Ceylon, the demand for Chinese tea was greatly reduced.

So great was the English love of tea that by the mid-eighteenth century, it was the cause of much apprehension. When tea first reached Europe, a curious belief spread that Chinese tea was draining the energies of people in the West. This was ironic in light of the fact that the British exported opium to China. Others felt that precious money and time were being squandered on the beverage. In a public letter, the writer and lexicographer Samuel Johnson responded vehemently to one such attack, describing himself as a hard, shameless tea drinker whose kettle scarcely ever had time to cool.

Homage to tea is found in the works of many other famous English writers and politicians, some of whom tout the drink as a panacea. I once read a rather pragmatic poem by William Gladstone that I found on the back of a menu at a tea shop in Portland, Oregon:

> *If you are cold, tea will warm you*
> *If you are heated, it will cool you*
> *If you are depressed, it will cheer you*
> *If you are excited, it will calm you.*

Before tea became a custom in English homes and tea shops, it was celebrated in London's famous—or, according to some, infamous—tea gardens, among which were Marylebone, Ranelagh, and Vauxhall. There, from breakfast time on, society was offered amusement and entertainment. After paying an admission fee, patrons could drink as much tea and eat as much bread and butter as they desired. During warm weather, friends and couples would meet and promenade along intricate gravel paths, enjoying the pavilions and arbors and watching fireworks in the evening. Some of the gardens had buildings large enough for dancing or concerts, at which Handel sometimes conducted.

The growing popularity of tea did not go unnoticed by the English government: taxes on it were raised, and the resulting higher prices tempted smugglers. Tea smuggling became so widespread that it actually reduced the cost for clever shoppers, who, by drinking smuggled tea at home, avoided the high taxes. In time, as more tea was consumed at home, fewer people went to the gardens for tea. However, the illegal trade allowed the practice of selling contaminated tea in order to bolster profits.

At first, it was suspected that dried ash, willow, and elder leaves were added to tea. Although these additives seem to have gone more or less unnoticed, the use of sawdust, gunpowder, and even dried sheep dung soon became apparent to even the most blasé drinker. These problems caused more people to turn to black tea; although the taste of green tea was preferred, its adulteration was becoming more evident. Soon England's parliament was faced with complaints that entire forests were being cut down to provide sawdust for adulterated tea. The whole absurdity led to the first laws against such adulteration in 1875.

The English usually add milk and sugar to black tea. Some believe that the idea of milk in tea came from Mongolians and other nomads who habitually and ritualistically mixed goat, camel, or cow milk into their tea. It seems more likely, however, that the English discovered that the heavier, darker tea was more palatable with milk. Besides, cold milk poured into a fragile porcelain cup before the addition of the scalding tea prevented cracking.

The Portuguese princess Catherine of Braganza, who married England's Charles II in 1662, is credited with introducing tea at court. Her dowry included a chest of green tea, which she preferred to take without milk or sugar. Edmund Waller wrote a birthday ode to her and to tea; it is thought to be the first English poem on the subject:

> *Venus (has) her Myrtle, Phoebus has his bays;*
> *Tea both excels, which she vouchsafes to praise.*
> *The best of Queens, and best of herbs, we owe*
> *To that bold nation, which the way did show*
> *To the fair region where the sun doth rise,*
> *Whose rich productions we so justly prize.*

The Muse's friend, tea does our fancy aid,
Repress those vapours which the head invade,
And keep the palace of the soul serene,
Fit on her birthday to salute the Queen. [19]

The custom of having tea for breakfast began with Queen Anne, who reigned from 1702 through 1714. She introduced the use of large silver teapots in place of the smaller ceramic ones from China. But it was Anna, seventh Duchess of Bedford, who in 1840 thought of the idea of afternoon tea. Perhaps she felt it would be a wonderful way to entertain and be entertained, or perhaps she saw it as a means of staving off that hungry feeling she sometimes had in the late afternoon. Whatever the reason, afternoon tea soon became a social challenge. Aristocrats and the moneyed class vied to see who could provide the most interesting hospitality, the fanciest pastries and music, and the most elegant service. This competition led to elaborate teapots, expensive china and liqueurs, fine claret, liveried footmen, and unpredictable conversation that might range from the weather and philosophy to the worst and wittiest gossip. It also led to "high" and "low" tea. For aristocrats, low tea became an opportunity to snack before an extravagant dinner. High tea was more a habit of the middle class and was often called "meat" tea, because meat, bread, and cheeses left over from substantial midday meals were often substituted for supper.

The Duchess of Bedford's innovation became so important to the English that it was soon claimed as a right by every worker. The tea break was relished and appreciated because it brought people together. Even until recently, workers went on strike over the issue of their tea breaks.

Afternoon tea offered a chance to pause, to enjoy a sense of peace during a possibly trying day. Such interludes might spark a romance, a reunion, or rich discourse. Whatever it inspired, teatime was, in a way, a Western counterpart to the more aesthetic Eastern ceremony.

[19] Ukers, page 133

Tea and...

In the United States

Tea was not particularly popular in the American colonies until the early 1700s. Eventually, though, it became something of a staple. It also became a famous problem when the English parliament retained the tax on tea after repealing taxes on other imports. The revenue was intended to help the financially embarrassed British East India Company, but the American colonists, feeling oppressed by more than taxes, revolted. The tax on tea was the catalyst for the historic Boston Tea Party on the night of December 16, 1773. Dressed as Mohawk Indians, colonists boarded trading ships and brazenly dumped 342 chests of tea into Boston Harbor. The revolution that followed affected the popularity of tea in the New World for more than a hundred years. Although tea was still consumed, it never gained the prominence it enjoyed in England and Canada until a new way of serving it was devised at the 1904 World's Fair in St. Louis.

The story goes that the government of India appropriated $75,000 to promote its exports at the fair by building a pavilion to exhibit tea, coffee, jute, and other products. Three thousand dollars was added to the fund by the Tea Association of India, which appointed Richard Blechynden to supervise its interests. The exotic pavilion was a sensational reproduction of the tomb of Etmad-Dowlah at Agra, complete with green domes and minarets. On the balcony overhanging the entire courtyard, white-clad, turbaned Indians gracefully offered free cups of steaming black tea to the passing crowds. Unfortunately, St. Louis was experiencing the hottest July in memory, and no one was at all interested in hot tea. In desperation, Blechynden thought of brewing stronger tea and adding ice to each serving. This "iced tea" was such a sensation that he carried his idea to other cities. His effort so spurred sales of black tea that it almost eliminated the importation of green tea from China, which had been the first choice of most Americans. Blechynden was apparently not the first to have had such an idea, however:

"the cooling and invigorating influences of iced tea" had been mentioned in a January 1880 issue of *American Punch*.

The tea bag, another American innovation, was also a chance creation. In order to interest his customers, Thomas Sullivan, a frugal New York City importer, had assorted teas sewn into small silk bags and sent out as samples. The outcome of this promotion was quite unexpected. One story has it that a recipient dropped one of the silk bags into hot water by accident and was delighted with the result. Another version holds that someone deliberately dipped one of the bags in hot water. Whatever the case, customers, much to Sullivan's surprise, clamored for more "tea bags."

The earliest known cultivation of tea in the United States was conducted near Summerville, South Carolina, from 1890 to 1915. Although the Southern climate was suitable, labor proved expensive, and because there was no tariff protection after 1903, the effort was all but abandoned. Today, some tea is still grown in that region; it is known as American Classic, "the only tea grown in America." The United States has now become the world's second largest importer of tea. By the 1990s, almost one hundred million tons were being used each year in teabags or processed into instant tea.

Today, more and more Americans are discovering the pleasures of tea. At hotels as well as at a growing number of small tea shops around the country, afternoon tea has become popular. Health conscious consumers and those cognizant of the other benefits of tea see it as an intelligent alternative to coffee. Although coffee is still the favored American drink, more than fifty billion cups of tea are enjoyed in the United States each year. There is even a trend toward Chinese teas, especially green and oolong varieties. Once found only in specialty shops or Chinese herbal-supply stores, Chinese tea is now more available in select markets.

This growing interest in tea in the United States—including the phenomenal interest in herbal beverages that began in the 1970s—might today surprise the turbaned Sikhs of the 1904 World's Fair, who never realized it was too hot for tea in St. Louis. It is a revival that might please Queen Anne or Samuel Johnson. But to Lu Yu or the venerable Rikyu, it would seem meaningless if it did not also renew an interest in ceremony—in the subtle social and personal implications of making and taking a bowl of tea.

3
Varieties of Tea

When touching on the history of tea and tea cultures around the world, it is always wise to remember that many of the legends and stories that have survived through centuries are colored in translation and interpretation. More factual and practical are the methods by which tea is processed, the many ways the beverage is enjoyed, as well as old and more recent studies of the health benefits of tea. Considering the time, effort, and expertise that go into making a thoughtful bowl or cup of tea, it is surprising how few people know much about its processing. Such knowledge helps one appreciate the different types of tea.

Tea can be divided into three general categories: unfermented green tea, semifermented oolong, and fully fermented black. Unfermented leaves can be sold as loose-leaf or, as in Japan, processed into powdered tea that is whisked directly in hot water, creating a frothy brew. During the Tang Dynasty in China, fermented leaves were pulverized into powder and whisked in hot water, but these did not dissolve as well as green, unfermented leaves. Instead, a dark residue of powdered tea settled at the bottom.

When I began sampling green tea, I learned that the best loose-leaf tea consists of small leaves. To the expert, large-leaf tea, although a dependable product, hardly compares in taste. Broken leaves, whether originally large or small, do produce good tea, but one cannot always be certain of the quality of what one is buying. The fannings (small particles) of processed leaves and the dust left over from sifting are sometimes added to teabag tea. Since fannings and dust are considered second rate, I find it better to purchase loose tea.

Unfermented (Green) Tea

Various brands of green tea that come from China and Japan offer a refreshing change from the more commonly used black teas. Green tea undergoes minimal processing and thus retains more of its vitamins and minerals; perhaps that is why I am partial to drinking it. In practicing the Japanese tea ceremony, I have developed a special fondness for powdered green tea whisked in a bowl, where its brilliant jade color and frothy scent seem to intensify the distinct taste.

Some of the finest powdered tea comes from the Uji district of Kyoto; however, it can be expensive and not easily obtained in the West. While I was in Japan in 1993, my curiosity about how powdered tea was processed led me to visit the Kyoto Prefectural Tea Industry Research Institute in Uji, a short journey by train from Kyoto. There, I learned that the cultivation and processing of *Gyokuro*, the institute's finest leaf tea, and *Tencha*, from which powdered tea is produced, are complicated and delicate procedures. The director of the institute at the time, Yoshiyuki Kawato, showed me how some of the tea plants are artificially shaded with reed matting or black plastic mesh as they grow. Two weeks before the first picking, rice straw is scattered over the shades in order to further regulate the intensity of the natural light. Organic fertilizers are then applied to the plants' roots. This increases the nitrogen content, helps produce more amino acids, and, according to Mr. Kawato, improves the taste of the tea leaves.

When the first crop is hand-harvested in May, only the first three to five leaves of each new shoot are picked. Depending on the quality of the leaves, there may be a second picking in July and a third in September. If these are not up to standard, further harvesting is suspended. To prevent fermentation, the leaves are steamed right after they are picked, then rolled and dried. As they dry, the leaves begin to curl, after which they are sorted and graded. *Gyokuro* leaves are packaged for loose-leaf, or brewed, tea. *Tencha* leaves, pulverized in stone grinders into a fine powder called *matcha*, are used for whisked tea.

Matcha is used mainly in the Japanese tea ceremony. Although not widely distributed outside Japan, certain grades may be found in select stores in Seattle, San Francisco, New York, Los Angeles, and Hawaii. (See Some Sources.) It is important to remember that, just as in the ninth century when Lu Yu wrote *The Classic of Tea*, the quality of the water still plays a significant role in making good tea. With this in mind, it is probably best to use filtered or bottled water.

John Blofeld, scholar and writer who was one of the first Westerners to write about the Chinese art of tea, informs us that to the Chinese connoisseur, the finest green tea grown anywhere comes from the peaks of the T'ieh Mu mountain range. Picked at the right time, "before each tender sprout has more than a single leaf," the tea is unsurpassed in color, aroma, and flavor. Its fragrance is so delicate that the leaves must be "skillfully packaged or kept in completely airtight tea caddies." (I have found it helpful to preserve the taste of powdered or leaf green teas by wrapping their containers in sealed plastic wraps and placing them in the refrigerator or freezer.) Blofeld warns against over-long steeping, which would "make the infusion bitter." However, if the tea is insufficiently steeped, "it will taste insipid." [1] Generally, it is best to bring the water to boil, keeping the temperature around 165° to 180°F.

Tea from the peaks of T'ieh Mu can also be found in the Chinatown districts of certain cities. [2] Offering a fine selection of tea for tasting in thimble-sized cups, these shops also provide an opportunity to ask about such exotic green teas as *Pai-Yun* ("white cloud"), *Jin-Chu* ("sun-poured"),

[1] Blofeld, page 73
[2] See Some Sources, page 203

or *Shou-Mei* ("old man's eyebrows"). Some of these shops also have a moderately priced to expensive selection of old and new Yixing teapots, a few of them from private collections. The Yixing teapot, in various sizes and designs, is a delightful vessel for serving green tea.

Semifermented (Oolong) Tea

Semifermented tea is produced by spreading the leaves and partially drying them for about twenty-four hours, after which they are put through a machine that rolls, twists, and breaks them. The larger pieces are sifted out and rerolled. The bruised leaves then begin to oxidize and are left to ferment for several hours. This process changes their color to a reddish-brown. Once the desired color is reached, the leaves are fired with hot air and dried, a process that stops their fermentation. Well-known oolong teas are generally 60 percent fermented. Oolong tea is best brewed with water just below the boiling point. Although oolong teas differ in taste from the lighter green tea, many people feel that they are just as flavorful and healthy. In Asia, it is said that the best oolong grows in the Wu-I mountains in Fukien, China. In the West, Formosa oolong from Taiwan is better known.

The color of semifermented tea lies somewhere between the green of unfermented tea and the dark copper of fully fermented, or black, tea. Because it is partially oxidized, semifermented tea is less bitter than green tea and not as strong as black varieties. There are three grades of semifermented tea: light, medium, and heavy.

Of the light semifermented teas, *Tikuanyin* is one of the most popular. Among the medium grades, *Chinchua* and *Dunting* are also an interesting change from regular black tea. When I sampled a fine *Tikuanyin* from Fujian at the Imperial Tea Court in San Francisco, I was told that it contained more caffeine than any other tea. However, I could discern no adverse effects from drinking a few cups brewed from the same leaves. I was later served such favorites as *Penfun* and *Mandarin*, so that I could try a heavier oolong.

Another rare grade of light semifermented tea is called white tea. Not actually considered an oolong, it is nevertheless treasured for its taste. I once tried a brand of white tea known as *Yin Zhen Bai Hou*, or "silver needle with white hair," at the Imperial Tea Court. Before brewing the tea, the waiter placed some leaves in a small dish for me to examine. Each was covered with a silvery fur. The leaves were then put in a fine ceramic cup, hot water was poured over them, and the cup was covered with its matching lid. When I removed the lid, an airy, flowery fragrance rose from the cup. The transparent golden brew was subtle and distinctly splendid.

Fully Fermented (Black) Tea

Black teas from China, also known as *congous*, are fully fermented. They are processed in much the same way as semifermented teas, except that the leaves are left to ferment longer.

One of the most favored black teas is *Keemun*, from Anhwei province. Ever since the 1700s, when the British East India Company began to import tea into England, *Keemun* has been fashionable because of its fragrance, color, and taste. Always popular, too, is *Lapsang Souchong*, whose famous smoky flavor results from drying the leaves over a cypress or pine fire.

Because black tea is usually graded by the size of the leaf rather than its quality, it may be difficult to know how to choose the best. *Pekoe* is the name given to top grades of black tea. Meaning "white hair" in Chinese, it originally referred to the youngest leaves, which were covered with down. Today, pekoe denotes a commercial grade of leaves that, after firing, are sifted through a certain size of mesh. In India, "broken orange pekoe" stands for the highest grade. The word "orange" has nothing to do with the flavor, but designates a certain size of leaf.

Most of the tea consumed in the West, apart from herbal teas, is black. Well-known blends range from English Breakfast, Darjeeling, and Earl Grey to the orange pekoes of Lipton and Tetley. These blends consist of teas from several regions that produce different tastes and strengths in different combinations. English Breakfast is a blend of Indian and Ceylon teas; Irish Breakfast is a blend of Assam and Ceylon teas. Ordinary brands may contain as many as twenty or thirty teas in the blend.

The flavor of tea comes from volatile oils, its stimulating properties from caffeine, and its astringency from the tannin content. Some varieties of black tea are flavored with citrus, mint, or oil of bergamot (an Italian citrus rind commonly used in the production of perfume, and added to Earl Grey tea). Other black teas are flavored with orange, or spices such as clove, allspice, or cinnamon. When subtly blended, these flavors can enhance the taste of a tea. Purists prefer that nothing be added to good tea, not even flowers such as rose or jasmine.

In India, England, Scotland, and elsewhere, the popular additives in black tea are milk and sugar. The host, before pouring tea sometimes asks if a guest would like "milk first" or "tea first." Each choice, it is believed, makes for a different taste. When I think of the myriad ways that tea is enjoyed around the world, I am reminded of some tea customs recounted by Terence O'Donnell, a late Oregon historian who once lived in Iran. There, as in other parts of the Middle East, tea is served in a slim glass, although the size of the glass apparently diminishes as one travels from the north to the south of the country.

Called *chai* (as in India), tea is as dear to the Iranians as coffee is to the French. A good glass of tea is said to require "*labdouz, labriz,* and *labsouz.*" *Labdouz* is when the tea is sweet enough to "sew the lips." *Labriz* is when the glass is filled to "the very brim." And *labsouz* is when the tea is hot enough to "burn the tongue."

Figure 7

Tea in a glass

Many people believe that tea, fermented or unfermented, loses flavor when it is decaffeinated. Caffeine is drawn out by treating tea leaves with pressurized carbon dioxide or with natural solvents; neither process, according to *Consumer Reports* (July 1992), leaves a harmful residue.

What makes an excellent cup of black tea? Experts agree that black tea should have a moderately sweet fragrance, a full body, and an average astringency and bitterness. These standards have made blending tea an art and, of course, a lucrative business. The more that we as consumers know about tea, the better we can discriminate and assess the various blends for our own satisfaction.

Like knowing wine, knowing tea, green or black, is a matter of experience. The more we sample, the better we can choose not only which teas to drink, but when and how to drink them. I find black tea best in the morning, green tea refreshing in early afternoon, and an herbal tea soothing after dinner. Just as I might prefer dry Riesling with sole, I might brew a strong pot of *Ty-Phoo* to accompany Jarlsberg cheese and whole-wheat sesame crackers late in the day. Where we drink tea also seems to affect the taste. When I spent time in Africa's Serengeti, I relished waking up to a hot cup of brisk, black Safari Tea, a pleasure that was sometimes accented by the bleating grunts of migrating wildebeests or the barking of zebras.

There are some occasions when we are served a superbly brewed cup of fine tea without the leisure to appreciate it, or, with time on our hands, we are faced with a cup of unpleasant weak tea. Although it is generally true that we get what we pay for, expensive tea does not guarantee satisfaction. A friend of mine, Jack, always examines the size, color, and texture of tea leaves in order to determine what kind of taste and aroma he can expect. To him such scrutiny is essential. Once his tea has brewed, he savors the color and aroma at the first sip as much as at the last, and then enjoys the remaining scent at the bottom of the cup. (Unlike tea "seers" I have known who dabble in fortune telling, Jack completely ignores any message left in the leaves.)

The taste of tea is influenced as much by region, climate, and soil conditions as it is by processing. The plant needs a sheltered, well-drained area, a fairly warm climate, and ample rainfall. The finest tea grows on mountain slopes, where the cool air allows the plant to mature best. In Asia, where tea is hand-harvested several times a year, the work is done mostly by women, who pluck the flush—the new leaves and buds—during periods of the tea plant's active growth. These flushes are then dried, rolled, heated, and fermented during processing.

Tea is often named after the region or town where it is grown or processed (such as Darjeeling), or for its color or grade. For instance, Gunpowder is finely balled young green tea, and Orange Pekoe, as we have seen, is black tea from the siftings of very fine, small leaves and buds. Teas are sometimes named capriciously but are usually given a distributor's name (Twinings or Bigelow) or named after a specific blend or method of processing.

It was once the practice to include the time of harvest on each package of tea, information that was always appreciated by the connoisseur. Today it is rare to find a harvest date, and it is usually found only on exclusive teas. When no harvest date is supplied, connoisseurs make a point of finding out the region or country where the tea was grown, as well as the grading system used. In China, the finest grade of leaves is picked before the Clear Light Festival, which begins in late March or early April when the leaves are small, delicate, and unusually aromatic. A similarly prized first harvest, the New Year's Tea, is treasured in Japan as much as nouveau wine. Respected plantations accept orders as early as a year beforehand for their crop.

Herbal Tea

As preventive medicine becomes an increasingly popular approach to health, we have returned to one of the oldest means of treating human ailments. More and more, the nutritional and medicinal properties of herbs are receiving the recognition they deserve. However, we need to define what we mean by "herbal tea," which is actually an infusion or beverage made by steeping fresh or dried plants other than the tea plant (*Camellia sinensis*) in boiling water. We should also make a distinction between herbal beverages and herbal medicine.

Since ancient times, humankind has benefitted from countless herbs, roots, and certain flowers prized for their medicinal properties, flavor, and scent. Originally, herbs were employed as spices and condiments. These uses led to herbal medicine, which was often based on the superficial resemblance of certain plants to specific human organs. For instance, wild ginger was said to be "good for the heart, since it looks like one." And the figwort, whose flower has a long throat, was given to those with neck and throat ailments. [3] Such classifications are all but ignored by modern medicine. The synthetic drugs used today have largely replaced the natural, inexpensive remedies found in an herb garden.

[3] Crellin and Philpott, page 14

Herbs not only add nutrition and spice to our food, but are also wholesome complements to our meals and wondrous additives to cosmetics and dyes. Before synthetics, the distinct tint of the indigo plant enriched the lives of millions by providing not only dyestuff for printing ink, but also one of the most attractive natural dyes for cloth. Long before Chinese tea was introduced to the West, Europeans favored peppermint, chamomile, and rosehip teas. Native Americans were sensitive to the pleasure and power of herbs and were careful to ask permission of a plant before picking it, giving thanks for its healing power and abundance.

As humankind's oldest remedies, herbs are steeped in folklore and myth. Even flower teas have their own fables and particular cures. Sweet chrysanthemum, for example, is an especially prized remedy in China. It is said that this beverage "cools" the body during the heat of summer. The ancient Chinese believed that drinking sweet chrysanthemum tea added to their longevity.

Blofeld recounts a tale from the Song dynasty in which the emperor Kao Tsung (1127–63) was so pleased with the performance of a troupe of palace dancers that he sent for their leader. As he chatted with her, she told him that her name was Chu (Chinese for "chrysanthemum") and that she was twenty-one years of age. An old eunuch who overheard her name thought he recognized her as the woman who had danced at the same court some forty years before—in which case Chu was sixty-one, and surely a witch! When the emperor was informed of this possibility, he demanded an explanation. Chu said that her family lived close to a stream called Chrysanthemum Rivulet, where they drew their water. It was therefore not surprising, she said, that "we mostly live to be over a hundred years old without any change in youthful ability and appearance." Others confirmed her story, thus clearing her of any suspicion of witchcraft. [4]

An herb connoisseur once served me chrysanthemum tea. Its flavor and scent were delightful. The small dried flowers opened fully in the pot as they

[4] Blofeld, page 90

steeped, and the golden color of the brew set off the simple cup beautifully. The tea was so light and delicate that, much to my host's satisfaction, I admitted that I could almost taste the fragrance.

In recent years, the pleasing taste of herbal teas and their blends has attracted those in search of caffeine-free beverages. During 1991 and 1992, more than $120 million was spent in U.S. supermarkets on herbal tea—and this figure does not include herbal tea sales in natural food stores and restaurants. [5] Although the demand for herbs continues to increase, their cultivation remains largely a cottage industry. Small herb farms do exist, but many of the herbs used for tea grow wild in fields and forests and on mountainsides. For the most part, they are picked by hand just as the buds appear on the plants. Depending on the type of herb and the intended use, the leaves, flowers, roots, bark, and seeds are all potential ingredients of herbal beverages.

Unlike black or other teas, herbs are processed naturally from the moment they are picked, ensuring their flavor and health benefits. Once harvested, they are tied in bundles and hung upside-down or laid out on screens to dry. In order to retain their natural colors and oils, which are so vital to flavor and quality, they must be dried immediately, either outside in the shade or indoors. They are then carefully packed in sacks or boxes for shipment to various tea makers, who clean, mill, and sift the herbs before blending them. Spices such as ginger, cinnamon, or cloves are added, or, for fruit-flavored teas, the leaves of the blackberry or raspberry. Every herb has its own distinctive taste and scent. Of all the varieties, chamomile and mint are the best known.

Some blends do far more than provide a satisfying or soothing beverage: they are also appreciated for their medicinal qualities. *Echinacea Immune Support* is a blend of echinacea root, elderberry, peppermint, lemongrass, cinnamon, fennel, spearmint and rose hips, which many use as an immune enhancer. *Throat Coat* and other teas from a package of *Cold Season*

[5] *University of California at Berkeley Wellness Letter,* January 1992

Sampler are effective aids to relieve colds. *Eater's Digest* is a marvelous after-dinner beverage, as is *Tension Tamer*, which lives up to its name. It's a blend of *Eleuthero ginseng*, ginger root, chamomile, and flavors of peppermint, cinnamon, and lemon. *St. John's Wort Tea* helps balance mood swings and ease depression. Blended with fennel seed, cinnamon bark, spearmint leaf, fenugreek seed, and other organically grown herbs, it is not only a pleasant drink but one way to seek equilibrium and harmony. A favorite of mine is a blend of African *rooibus* (red bush) and rose hips. With honey, it is a pleasant beverage at any hour of the day and, when ice is added, is especially refreshing in summer.

Like other teas, herbal tea is an acquired taste. Its success as a flavorful brew depends on the blend, though some believe that our physical and emotional states also influence our enjoyment. For instance, chamomile is better when one needs something calming; after a heavy meal, ginger is beneficial to the digestion. At work—or anytime, for that matter—mint tea can be taken. Deng Ming-Dao, in his book *Scholar Warrior*, mentions that mint tea is pungent and cooling: "It enters the lung and liver meridians and dispels wind and heat and cleanses the throat." [6] But there is always the need to be sensible. I was once served a glass of peppermint tea in a small perfume shop in Cairo's colorful Khan el Khalili bazaar. Heady from samples of musk and amber, and romanced by the sights and sounds of the busy stalls, I drank the hot, highly sweetened beverage. I found it instantly reviving, but wondered later when I felt somewhat depleted, whether it was the sugar, more than the mint, that had provided the energy rush.

With herbal brews, it is wise to be cautious. Herbs are powerful; they are, after all, used as medicine. As infusions, tinctures, decoctions, or extracts, herbs should not be taken indiscriminately or in large doses. Some, such as nightshade, are poisonous unless buffered with other herbs. Comfrey can be healing when taken prudently, but large quantities can affect the liver; lobelia can induce vomiting; and no one has fully documented what the

[6] Deng Ming-Dao, page 173

long-term effects of the raspberry leaf might be. In summary, the blending of herbal tea and remedies is best left to experts. For the novice, advice is sometimes available at natural food and Chinese herb stores and in numerous books on the subject in all major bookstores.

Tea and Health

The first cup caresses my dry lips and throat,
The second shatters the walls of my lonely sadness,
The third searches the dry rivulets of my soul to find
the stories of five thousand scrolls.
With the fourth the pain of past injustice vanishes
through my pores.
The fifth purifies my flesh and bone.
With the sixth I am in touch with the immortals.
The seventh gives such pleasure I can hardly bear.

Lu Tong [7]

The healing qualities of tea were appreciated and respected throughout the poet Lu Tong's ninth century China. The Chinese recognized the revitalizing properties of tea through their early practice of herbal medicine, dating back to 3000 B.C., when the Divine Cultivator, Shen Nong, created remedies with herbal teas. Hua Tuo, the third century surgeon who discovered anesthesia, believed that drinking tea increased alertness and concentration. According to Okakura, in Chinese medicine tea was prized for "relieving fatigue, delighting the soul, strengthening the will," and "repairing the eyesight." [8]

Although the drinking of tea did not become popular until the Tang dynasty (618–907), its health-giving qualities were already known among

[7] Chow and Kramer, page xiii
[8] Okakura, page 22

nomadic peoples on China's northern and western frontiers. Mongols, Turks, Tartars, and Tibetans, who subsisted mainly on meat and milk products, found that the frequent drinking of tea helped cure certain ills resulting from a lack of fruits and vegetables in their diet.

In the West, modern medicine is beginning to acknowledge what the East discovered long ago: tea drinking promotes good health. A six-ounce cup of regular tea has only two calories and contains several trace minerals, as well as vitamins B and C—giving it a place in a healthy diet and making it preferable to coffee if sugar is kept to a minimum.

Whether hot or iced, tea quenches thirst and, with a little lemon juice, increases vitamin intake. In addition, tea freshens the breath, aids the digestive process, sensitizes the palate, improves the appetite, and stimulates blood circulation. All teas are diuretics to some extent; they help stimulate the kidneys and other organs. The calming effects of tea can ease hypertension, and taking an aspirin with caffeinated tea increases the aspirin's effectiveness.

Some of the ingredients in tea may even help ward off cancer. In the *University of California at Berkeley Wellness Letter* for January 1992, scientists speculated about the antioxidants in a cup of tea. A class of chemicals known as polyphenols, which give tea its astringent taste and are most plentiful in green tea, may act like antioxidants, working at our body's molecular level and, some scientists believe, fighting against "highly reactive" oxygen molecules known as free radicals. Free radicals are thought to damage basic cell structure, thereby contributing to the development of diseases such as cancer. Because the scientists' study was done exclusively on animals, it cannot be considered conclusive. However, green tea "did seem to prevent tumor formation in rats. Animals fed green tea were less likely to develop skin cancer, even when exposed to enough ultraviolet radiation to promote it. Studies of rats also showed that green tea could lower blood cholesterol."

Adding milk or lemon to a cup of tea makes it even more nutritious. In a back issue of *American Health* (January/February 1990), writing about a "proper cup," Robert Barnett found that with lemon in tea, "both iron and calcium may become more available . . . the Vitamin C in lemons may make the minerals more soluble and easier to absorb. And adding milk adds calcium." Tea has a generous amount of fluoride — especially green tea, which has "twice as much per cup as black," helping prevent tooth decay. The tannin (a rich source of fluorides) in tea is believed to inhibit the formation of dental plaque. Studies of English schoolchildren have confirmed this finding, showing that those who drank tea had fewer cavities than those who did not.

Tannins may have other benefits. Barnett reminds us that they are "antibacterial, antiviral and they treat diarrhea." It is the tannic acid in tea that deepens its color and gives it an astringent taste; the more tannin in tea, the more bitter the taste. To reduce the bitterness, Barnett suggests that when we brew a pot of tea, we should bring cold water to a boil and then pour it over the tea leaves. Covering the pot immediately helps to keep the water at "that ideal just-below-boiling point as the tea steeps." Like many tea lovers, Barnett believes that a china, glass, or ceramic pot makes a better cup of tea, pointing out that metal combines with tannins and makes the tea "taste even more bitter." [9]

The Caffeine Controversy

Unlike the chorus in *Through the Looking Glass*, who suggested that we "put cats in the coffee, and mice in the tea," we do have concerns about what constitutes a cup of tea, and in particular about how much caffeine it contains. Garth Clark, writing in *The Eccentric Teapot*, estimates that an average cup of black tea has "about a quarter of the caffeine found in a cup of coffee while the same amount of oolong or green tea has even less." But experts agree that, whatever the amount, the caffeine in tea can be beneficial. It is known to cause the constriction of cerebral arteries, which may prove useful in treating vascular headaches. Unlike the caffeine in coffee, caffeine in

[9] Barnett, page 95

tea is not associated with coronary disease. [10] It is also believed that the caffeine in coffee is released differently from that in tea. Coffee gives up caffeine immediately. This release creates "a jolt of energy, which is then followed by an equally rapid decline, but tea, its devotees insist, releases its caffeine in a gentler, soothing arc." [11]

However caffeine is released, it is important to know its advantages and disadvantages. A well-known stimulant, caffeine boosts alertness, lifts spirits, and relieves mild depression in many people. However, in *Holistic Medicine* Kenneth Pelletier warns against using caffeine as "a self-medication for depression," and points out that the side effects of excessive caffeine include "irritability, dizziness, frequent urination, and free-floating anxiety." [12] Athletes who use caffeine to increase endurance run the risk of severe withdrawal symptoms; this is also true for other heavy coffee drinkers. The discerning lover of tea might begin with the wisdom of occasionally avoiding all caffeinated beverages—perhaps for one day a week, in order to allow the body to be free of caffeine.

When tea is consumed in moderation, its caffeine acts as a catalyst in refining our awareness of the moment. Lu Yu believed that tea *is* moderation in itself, that its *yin* and *yang* qualities of relaxation and alertness are experienced harmoniously. No more than three cups of tea should ever be taken at one time, and one should sip each as if it were life itself, according to Lu Yu. With moderation in mind, we can experience ceremony and see how the Way of Tea opens a door to tranquility.

[10] Barnett, page 94
[11] Clark, page 24
[12] Pelletier, page 157

4
The Experience of Tranquility

Whether simple or elaborate, rituals influence our behavior and the directions our lives may take. They enable us to deal with transitions as well as with the highs and lows of existence. The spirit of ritual is woven throughout the conscious and unconscious levels of our relationships and is an essential part of everyday life. We carry out rituals when we eat, dress, and greet one another with a handshake; when we bow or tip our hats. In both form and content, ritual enriches art, literature, dance, and music—and, above all, it illuminates myth.

The design and mood of certain rituals, from taking tea to attending mass, create a kind of theater—not for entertainment, but for the staging of finer experience. As theater, ceremony creates a special climate in which we explore the subtler dimensions of existence, such as the unspoken communication that occurs in tea rooms—especially with others who may not speak our language. Too often we are not cognizant of such dimensions, or have no way to appreciate them when we discover them. Such discoveries are complex, and our awareness of them depends on mastery of a style of life and our concern with spiritual renewal.

In modern life there is not enough serious use of ritual and ceremony. The mythologist Joseph Campbell warned us in 1949 that within our so-called

Figure 8

Guest drinking tea

progressive society, "every last vestige of the ancient human heritage of ritual, morality, and art" was in full decay. The democratic ideal of the "self-determining individual, the invention of the power-driven machine, and the development of the scientific method of research, have so transformed human life that the long-inherited, timeless universe of symbols has collapsed. . . . The spell of the past, the bondage of tradition, was shattered with sure and mighty strokes." [1] Because of Protestantism and its simplification of ritual, some elements of American democracy and humanism rebel against any group or ceremony that might emphasize hierarchy.

Campbell helps us realize that now, more than ever, we need to heed the rituals and ceremonies that nurture our lives and our spirits. While the spell of the past has been fractured, freeing us in countless ways from the bondage of dated belief systems and superstitions, there are aspects of

[1] Campbell, page 387

tradition that we cannot afford to lose. As we enter the twenty-first century, we can benefit by drawing upon cultural symbols that enhance our sensitivity and enrich our lives—a mandala that calms the mind, a candle that lifts the spirit, a flower in a bamboo vase that evokes the natural world. These symbols endure because they offer inspiration; like ceremony, they give our lives form, order, and meaning.

As tea became a culturally vested part of Western life, it never evolved into a spiritual practice as it had in the East. Jamie Shalleck notes that the taste for tea in China developed "hand in hand with artistic connoisseurship," whereas in the West "it came as a fad and a medical curiosity, not as a work of art." [2] The idea of monastic simplicity may have influenced tea in China and Japan, but afternoon tea in Holland and England was a more gregarious pastime that featured elaborate food and flowers and proper dress and etiquette, rather than spiritual renewal. Even today, taking tea in the West is appreciated as much for the opportunity to socialize as for its ritualistic or refreshment values.

Before I began to practice the Japanese tea ceremony, I sometimes gave Western-style tea parties for women in my home. I served various appetizers and cakes with several different fermented teas, as well as an herbal brew for those who preferred it. At first I invited only a few people, but I was surprised to find out how many of my friends were interested in this kind of gathering. I once gave a party for fifteen, which soon became unmanageable in the limited space of my living room. Everyone seemed to enjoy themselves, but I felt that something was missing.

Once I began my study of the Japanese tea ceremony and discovered the origins of tea, I realized that what had been missing was a sense of purpose. The tea ceremony has clear purposes: a unity with nature, peace, and the promotion of the ideals of art and beauty. Introducing Lu Yu's *The Classic of Tea*, Francis Carpenter says that ritual to the Chinese was one step "toward freedom and not a retreat." The "imposition of outward forms"

[2] Shalleck, page v

was the first step toward self-discipline. In turn, self-discipline was the first step toward self-realization. The Chinese thus believed that form, ritual, and rules supported rather than hindered freedom. Without this support, the undisciplined mind could not attain freedom. Such training of the individual was not intended as a "sterile imitation of past forms," but as "a means of finding what past sages had found or at least searched for."

Throughout his life, Lu Yu insisted on the importance of ritual in tea. He believed it was "another way of celebrating another act of living," and felt that ritual was so important to tea that if even one implement was missing, it was better not to have tea at all. One must know not only when the tea was plucked and the source of the water, but also who should be invited to share the tea. "Should a guest be missing," he warned, "the quality of the tea must be such as to atone." No detail should be neglected. In addition to using proper ingredients (the tea had to be of a delicate quality and the water had to be pure), keen attention had to be paid to the selection and preparation of the tea bowl as well as the accompanying utensils. All had to have "an inner harmony expressed in the outward form." If the host found the color of a certain utensil inappropriate, it should "be banished from the equipage," no matter how rare or expensive it might be. [3]

Chanoyu: Traditional Japanese Tea Ceremony

Realizing the spiritual aspects of the tea ritual requires a great deal of patience, combined with gracious dexterity and a genuine commitment to practice. Although learning the technique of tea preparation is important, training the mind to be peaceful and serene is considered most essential. Ritual calls upon the practitioner not only to study, but also to contemplate.

The late Zen master Taizan Maezumi Roshi clearly indicates the importance of ceremony in a transcript from a talk he gave at the Los Angeles City Center in March of 1994. The word "ceremony," he reminds us, is related

[3] Lu Yu, pages 5, 6, 7

to the Latin word *cura*, to heal. Thus ceremony is a way to heal, to be made "whole and healthy. . . . When our actions are done within a definite form, it becomes a ceremony." He further said that ceremonial action, "in order to be done well, has as its base the individual as the group—as the society, the country, or the world. By taking care of things in a ceremonial way, we become unified. We avoid our own self-centered interests."

When the tea ceremony was introduced in the Zen temple, it became as structured as Zen life. Many aspects of *chanoyu* are similar to Zen practice because they both deal with correct posture, concentration, and breathing—all requirements of Zen meditation. In *chanoyu*, as in Zen practice, breathing is centered in the lower abdomen—the *hara*, or "center" of the body. Once the abdominal muscles are trained, breathing becomes deep and natural, bracing and invigorating the body and mind.

When Rikyu refined *chanoyu* in the sixteenth century, he encouraged aesthetic simplicity, or *wabi*—the attitude that expresses communion between a human being and nature. Rikyu's concept of *wabi* had a profound effect on the Way of Tea, as did his four principles and seven major rules. These principles and rules govern the ceremony and encourage a natural flow of movement. Each movement is exact, and the entire procedure reflects an elegance and economy of action. The four principles are

Harmony (*Wa*)

Harmony comes from the focused efforts of both host and guest to create an atmosphere of peace. This twin effort reflects kind consideration and produces a mutual attention that is as natural as the rhythm of nature. The host thoughtfully chooses a theme that sets the tone for the occasion, often one that is inspired by the season or time of day and is, in turn, reflected in the calligraphy, flowers, utensils, and the food. The harmony of all these elements encourages contemplation, humility, and moderation; it also heightens a sense of the passing moment.

Respect (*Kei*)

Respect refines the exchanges that take place between host and guest and among the guests. It encourages everyone to look within and discover the inherent bond between themselves and their environment—their dependence on one another and the earth upon which they live.

Purity (*Sei*)

Purity begins with the physical surroundings of the tea gathering. The garden is carefully weeded, swept, and watered before the guests arrive. Purity is also expressed in the care given the tea room and utensils before, during, and after the ritual. This preparation and attention to detail establishes an essential order that centers the host. The cleansing of certain utensils with a silk cloth in front of the guests is a further attempt on the host's part to clear her or his mind of extraneous thoughts. The guests, too, perform gestures of purity, such as rinsing the hands and mouth before entering the tea house. The aesthetic ideals of *chanoyu*, which emphasize the use of simple, natural materials and the avoidance of ostentation, lend another aspect of purity to the gathering.

Tranquility (*Jaku*)

Tranquility is, in one sense, a coming together of harmony, respect, and purity. When the first three elements are present, tranquility is experienced, and when all four are unified, the experience is memorable.

Soshitsu Sen, fifteenth-generation tea master of the Urasenke *chanoyu* lineage, who is a descendant of Rikyu and lives in Japan, tells the story of a disciple who once asked Rikyu what were the most important things that needed to be kept in mind at a tea gathering. Rikyu's answer defined his seven rules:

- Make a delicious bowl of tea
- Lay the charcoal so that it heats the water
- Arrange the flowers as they are in the field
- In summer suggest coolness; in winter, warmth
- Do everything ahead of time
- Prepare for rain
- Above all, give those with whom you find yourself every consideration

The disciple, expecting something greater from the master—perhaps an exalted secret—seemed dissatisfied. What Rikyu had said sounded simplistic, and the disciple began to complain that he already knew as much. "Then," said Rikyu, "if you can host a tea gathering without deviating from any of the rules I have just stated, I will become your disciple." [4]

The seven rules that guide tea practitioners sound practical and simple. But like the four principles, they transcend what they state:

- Making a delicious bowl of tea goes beyond the quality of the brand. It rises above the elegance of the utensils and even the expertise or dexterity of the host. It involves the heart and the sincerity with which the host serves tea.

- A host has to be skillful in laying the charcoal so that it heats the water. To bring out the best in the tea, the temperature of the water must be just right. Keen attention is given to the preparation and form of the ash in a brazier or hearth. Steady

[4] Soshitsu Sen XV, *Tea Life, Tea Mind*, pages 30, 31

heat is produced by carefully arranging each piece of charcoal on the ash as prescribed.

⤷ The art of placing flowers for a tea ceremony is unlike *ikebana*—a formal flower arrangement that takes time and permits the use of tools and other accessories for props. In tea, flowers are displayed naturally, as they might be seen in the fields. They are picked, their stems cut under cold water, and arranged in the hands. The flowers are then placed with one breath—forward facing in a vase, while they are still fresh. However, this is never done haphazardly. Rather, the inherent beauty of each flower is revealed so that its precious quality is distinguished and its delicacy becomes a symbol of the transience of life. Buds, rather than fully opened flowers, are used to suggest this transience and reflect the preciousness of the time that guests will spend with one another.

⤷ To suggest coolness in summer and warmth in winter, the host can create subtle effects that bring comfort to the guests. The imagination is needed as much as technology and natural resources. In summer, the brazier is placed in a far corner, away from guests. Certain artifacts are used to reflect coolness, such as a shallow tea bowl or a dampened water jar in summer. Sometimes a large leaf sprinkled with dew makes a perfect lid for the water jar. In winter, a fire in the sunken hearth facing the guests will warm the tea room, and in December a mid-morning or early afternoon tea can be scheduled after the sun has warmed the earth.

⤷ Doing everything ahead of time shows respect and consideration on the part of both host and guest. If all participants allow enough time and are prepared for any unexpected circumstances, composure and freedom are ensured.

Preparing for rain is a wise precaution that encourages a willingness to confront the unexpected. Making sure umbrellas are at hand or remembering an aged guest in wet weather is the essence of generosity.

In discovering the Way of Tea, it is essential to give every consideration to one's guest. This seventh rule cultivates thoughtfulness and inhibits class distinction both in and outside the tea house.

Figure 9

Japanese tea house

There is no performer or audience in a tea gathering, only "an interaction of human beings" who, through sympathetic intention, become one. Because guests are in harmony with one another, they merge "into a single entity that transcends their respective roles." [5]

Early on in my study of tea, I was moved when I heard the story about Rikyu being invited to have tea in the home of a tea producer. The man was overjoyed when Rikyu accepted his invitation. As the host was preparing tea, however, he became so excited that his hands trembled; he inadvertently dropped the tea scoop, then knocked over the whisk positioned on the tatami mat. One of Rikyu's students who had accompanied him to the tea witnessed this performance. Nevertheless, Rikyu thoroughly enjoyed his bowl of tea, and told the host that it was among the finest he had tasted. On the way home, the student asked Rikyu why he was so impressed by such a clumsy performance. Rikyu was quick to answer that the man had not invited him with the idea of displaying his skill, but had simply wanted to make a bowl of tea with his whole heart. The host's sincerity was what moved Rikyu.

Of the many schools of tea in Japan today, three were founded by sons of Rikyu's grandson, Sotan. One, *Omotesenke,* was established in the front part of the family home. Another, *Urasenke,* was in the back part of the same house. And the third, *Mushanokojisenke,* was named after the street. In all three schools, although certain procedures have been refined, Rikyu's principles and rules remain intact. Students are expected to master procedures with the body as much as with the mind, so that their bodies respond spontaneously. In this way they are, in time, able to deal with any circumstance or mishap with composure and ingenuity. The roles of guest and host are studied with equal attention, for it is believed that the needs of one are inherent in those of the other.

Tanaka tells us that the preparation of tea requires three basic elements: arrangement, purification, and calmness of mind. Arrangement includes the way utensils are carried into the tea room and placed so that "their beauty

[5] Soshitsu Sen XV, *Tea Life, Tea Mind,* page 40

can be admired from many different angles." Using a silk cloth to cleanse the caddy and tea scoop, the host also purifies the mind and heart, and if these movements are done "without pretension," they bring on a "feeling of peace and tranquility in the guests." [6] All students endeavor to achieve this.

Tea houses, tea rooms, and tea gardens differ in size and location. The experience of a traditional ceremony also varies according to place, time, and whether or not a meal is included. In a simple ceremony, the serving of sweets and powdered thin tea, or *usucha*, may take up to an hour. *Koicha*, or thick tea, is usually part of a full-length gathering called *chaji* when a charcoal fire is laid in the hearth or brazier and food is brought to the guests. Considered the essence of hospitality, a *chaji* ritual can last as long as four hours or more. *Sake*, a Japanese rice wine, is served with the meal, *kaiseki*, which is a light repast. (The term *kaiseki* originated in Zen temples. It represents a time of intensive training when monks, who were meditating and fasting for hours, sometimes placed small heated stones on the folds of their robes close to their stomachs to relieve feelings of hunger.) At the end of the meal the guests retire for a short break before reentering the tea room. The mood is quiet, more intimate, as a bowl of thick tea is made and shared in communion among the guests. The tone lightens as the fire is replenished and *usucha* (thin tea) is served to each guest.

Birthdays, holidays, or events such as the New Year or the opening of the sunken hearth season (*Robiraki*) in early November, are celebrated by traditional tea gatherings. A gathering may be inspired by the dawn, a full moon, the return of a loved one, or perhaps a meeting of close friends or relatives—even a memorial. I have attended many such gatherings throughout the years, and in small and large ways each offered a new experience: a rare bowl, a singular silence, an unexpected glint of sunlight, or a peculiar song that rose out of the kettle's steam. An iron kettle imparts a distinct sound when containing boiling water. Aside from the temperature of the water, the sound varies with the size and shape of the kettle. The steam

[6] Tanaka, page 90

rising from an iron kettle resembles the sighing of wind in the pines. Such pleasures offered in a tea gathering illustrate *ichigo ichie*, a popular tea phrase that translates as "one time, one meeting" or, as a Zen master once said, "unprecedented, unrepeatable." This means that each gathering, such as it is, is an experience that will never occur again.

Invitation to a Japanese Tea Ceremony

Most practitioners agree that the primary consideration in any tea gathering is its preparation. When I receive an invitation, I am reminded that, just as the host prepares for the guest, the guest must also prepare. This dual effort promises the kind of removal out of which tranquility may be realized—perhaps not in a practitioner's earliest gatherings, but eventually, with experience and patience.

A particular gathering in a tea house or tea room may involve only a bowl of thin tea and some sweets. I sometimes prefer this to the *chaji*, because a simple gathering can be more spontaneous. An intimate bowl of tea with a friend always seems to enliven an ordinary day. However, the formal gathering of a *chaji* is special, since it provides an opportunity to watch the host lay charcoal in a brazier or sunken hearth and allows the guests to relish a fine meal, *sake* wine, and thick and thin tea.

Invitations to tea gatherings are designed by the host and give a hint of the theme chosen for the gathering. Whether the invitation is creative, poetic, or simple, the personal hand of the host is always involved. Receiving it, the guest responds with respect. The invitation I received to a particular afternoon thin-tea gathering reflected the season, spring, and the occasion, the celebration of a friend's birthday. It was from Yuko, a teacher of *chanoyu* who practices the *Urasenke* tradition of tea, and it offered the following poem beneath a small illustration of cherry blossoms:

Spring!
Dressed with plum and cherry
The garden waits.

The invitation also gave the time and place, and the names of other guests who had been invited. This four o'clock gathering was for Jack, a fellow student of tea ceremony. This amused me, because Jack did not celebrate birthdays and had agreed to come only because he knew we all understood that he loved tea and that the tea would be his only gift.

In any *chanoyu* gathering with more than one guest, someone is always designated as the first or lead guest, the *shokyaku,* and the last guest, or *otsume.* The first guest's role is to contact the other guests reminding them of what they are required to bring, the order in which they will enter the tea room, and the monetary donation, which is usually obligatory for a Japanese tea gathering. The donation, or *mizuya mimai,* is sometimes written on the invitation; but if not, a guest reflects on it and responds with the appropriate check or cash tucked in a card or simply placed in an envelope with his or her name written discretely at the bottom of the envelope. The *shokyaku* collects these from the guests when they meet at the gathering and presents the donations to the host at an appropriate time, or may place the envelopes in a discreet corner of the alcove of the tea room.

The conversation in a tea room begins with the host and first guest, so the first guest acts as spokesperson for the others. The role of the *shokyaku* is a sensitive one and influences the tone of the gathering. The first guest is expected to be aware of procedures and etiquette and of the right moment for conversation. The last guest (one who is as experienced as the first guest) anticipates every detail of ceremony and any need that may arise.

I drove through Portland's Friday afternoon traffic and arrived at the home of Yuko, who lives in the hills of Lake Oswego. In the Japanese tea ceremony, the person making the tea is always referred to as *teishu*—the host, whether man or woman.

After I got out of my car, I paused for a moment in order to clear my mind. I noticed Jack's car and that of Anne, another friend and tea student. Beyond the side of the house, Yuko's garden and tea house seemed refreshingly quiet under two tall pines. Mentally brushing off the stress of traffic, I slowly walked around the corner to the familiar porch attached to Yuko's main house. I greeted Jack, who opened the door and, leaving my shoes outside, put my coat and handbag in a hall closet before joining him and Anne. I had with me the customary pair of white socks worn in the tea room, and I carried a *fukusabasami*, or small fabric purse. In this was a fan, or *sensu*, a silk napkin, or *fukusa*, another piece of folded fabric called the *kobukusa*, a fold of small white papers called *kaishi*, used for holding sweets, a sweets pick, and a handkerchief for drying my hands after using the water basin. Sitting down on a chair in the living room, I put on my socks and agreed with Anne that we were all lucky to have a host with a garden and separate tea house. Jack nodded, adding that Yuko not only taught tea but, by her own humble admission, never stopped learning. "The keenest student in a class is always the teacher," she would say to us. Since he was designated as *shokyaku* (first guest), Jack mentioned that I would follow him into the tea house, and then Anne, who was *otsume* (last guest), would follow me. We would be served tea in the same order.

Because we would be sitting on tatami mats in the tea room, we were all dressed comfortably though somewhat formally. I wore a mid-calf knit dress with sleeves and Jack had on a sports jacket and his worn but favorite green tie. Tall and slender, Anne looked striking in her yellow patterned kimono, which is the usual attire for a traditional Japanese tea gathering. (Since an appropriate kimono for both men and women is usually difficult for Westerners to acquire and assemble, it is not absolutely necessary for them to wear one.) Following custom, we had put away our watches and rings so that they would not clang against a bowl or mar a utensil; and neither Anne nor I had used perfume or lotion as any scent would have interfered with the incense, the fragrance of the tatami, and the tea itself.

Just before four, Jack said it was time to go to the outdoor arbor where we would be greeted by Yuko. We each slipped on a pair of thonged sandals that Yuko had left for us and walked out into the garden toward the tea house. I tucked my fabric pouch into the side pocket of my dress, and in my right hand held the small folded fan, the *sensu*, which is never used to fan oneself. As a symbol of humility and respect, the fan is kept folded and placed on the floor in the tea room in front of oneself when greeting a teacher, host, or another guest. When placed behind oneself, it delineates one's own space. At certain times the guest places the fan between himself or herself and an object that is being faced and viewed in appreciation.

It was a fine day with a marvelous spring sky overhead and air fragrant with pine. We walked past a flowering plum tree and approached a small wooden bridge over a shallow pond. A pair of spotted carp were circling lazily below the surface; occasionally one would dart upward, then drift down again into the still water. From the peace of the pond, we strolled toward the secluded, L-shaped arbor in a corner of Yuko's garden—the outdoor *machiai*, where her guests wait for her. There, we relaxed on the round straw cushions of the wooden bench and waited for our host. I felt sheltered and quiet. The faint rumble of afternoon traffic was easily dissolved by the whisper of a breeze through the nearby pines. Breathing the garden air, I slowly began the ritual of letting go, of allowing my mind to settle on my breath and what was present before me.

Soon Yuko emerged from the tea house and approached, elegant in a pale lilac kimono. Tucked into her dark *obi* (the wide brocade sash about her waist) was a *fukusa*, the symbol of the host, a red silk cloth folded in a triangle. She greeted us with a gentle bow, a silent welcome. We three rose and bowed with her, holding our fans in our right hand. She then turned and walked back to the tea house. When she was inside, Jack bowed to me, excused himself, and moved toward the tea house. I then bowed to Anne and preceded her down the path.

One by one we passed through a small wicker gate, stepping on the wet stones that guided us through the design of the tea house's inner garden, called the *roji*, or "dewy path." To the Zen Buddhist, this is a path on which "one sheds worldly dust"—all the attachments and confusion that cloud the mind and heart. A tea garden is usually a small area built to resemble a path in the mountains with trees, bushes, a few flowering plants, and nothing extra. Swept clean and well-weeded, Yuko's tea garden looked refreshing. The stone path and azalea bushes had been sprinkled with water before our arrival, and their scent led us to the *tsukubai*, a stone basin brimming with water. Across its rim lay a bamboo ladle. As Jack had done before me, I knelt down and picked up the ladle. Taking a scoop of water, I rinsed one hand, then the other, discarding the water outside the basin on

Figure 10

Guest scooping water

surrounding pebbles. Taking another ladleful, I rinsed my mouth with the chilly water. Then, tilting the ladle up, I let the excess water run down the handle to rinse it before setting it over the basin again. I have always loved this simple cleansing. Part of the principle of purity, it is a wonderful way to symbolically wash the dust of the world from the mind. Leaving the water basin, I approached the tea house.

A natural, unpretentious structure, the Japanese tea house, built a foot or two off the ground, is usually part of a small garden. The tiled or thatched roof, gray stucco walls, wooden beams, and low windows offer an air of serenity and seclusion. The precision with which complex joints are secured reflects the builder's deep understanding of the integrity of wood.

Though traditional tea houses vary in size and appointments, each has a tatami mat tea room with an alcove and generally an adjoining room called a *mizuya*, which is a preparation area used by the host. The *mizuya* has its own sliding door into the tea room. The guest's entrance is a floor-level door about two feet square, allowing just enough space for guests to slide in on their hands and knees. This way of entry encourages humility, respect, and a feeling of equality. The favored tea room is not more than four and a half tatami mats in size—approximately ten feet square—and the beauty of its simplicity is heightened by aged wood, papered windows, and an alcove that welcomes shadows.

Yuko's tea house was traditional, although it had the modern amenities of running water and electricity. The small sliding door for the guests had been left ajar. Once Jack went through this door, I stepped onto the wide stone just below the entry and stooped in order to place my folded fan inside. As I looked in, the first thing I noticed was the alcove, or *tokonoma*, in which a long scroll hung. On the floor below the scroll was a slim celedon vase holding a single flower.

An iron kettle steamed in a sunken hearth to my left, and nearby stood a small round lacquered stand (*marijuko tana*) with utensils. Savoring the scene, I paused again. Before sliding into the tea room on my hands and knees, I slipped my sandals off and left them on the wide stone. The smooth tatami surface made a modest entrance easy. Then I turned and reached outside to place my sandals upright against the edge of the stone, just as Jack had before me. Following tradition, I picked up my fan and walked over to the alcove, where, facing the scroll, I knelt, placed the fan in front of me, and bowed. On the scroll was brushed Japanese calligraphy that Yuko had chosen to define Jack's only wish for his birthday—*jaku*, which means "tranquility." The white camellia in the vase had opened slightly, and it glistened in the light slanting through a side window. When Anne entered the

Figure 11

Alcove with scroll and flowers

Photo by Ted Bagley

room, she closed the entry door firmly. This was the way to alert our host that we were all in the tea room.

Picking up my fan, I rose and walked over to the small hearth, taking care not to step on any of the black cloth borders of the mats as I crossed from one to the other—something the guest (or host) has to be mindful of in accordance with the appropriate movements on tatami. The heavy kettle sat on a trivet over glowing charcoal. Like Jack, I took a moment to listen to the steam, enjoying the incense, a blend of fragrant wood particles and spices that rose off the bed of ash. On the small utensil stand was a red-lacquered tea caddy, the *natsume*, containing the powdered tea; on the shelf below sat a cold-water jar, the *mizusashi*, a ceramic piece with a light blue-green glaze and matching lid.

Once we were all seated, the silence in the room signaled Yuko. She slid open the door of the preparation room and carried in a tray of traditional Japanese sweets. Kneeling in front of Jack, she placed the tray on the tatami. She bowed and rose to leave the tea room, gliding across the mats. With the rustle of her kimono, her movements reminded me of something my first teacher, Nakamura Sensei, once claimed: that she could judge the attitude of a host by the sound made as she or he walked across the tea room.

Figure 12

**Lacquered stand
with tea utensils**

When the door of the preparation room slid open again, Yuko sat at its entrance with a blue tea bowl by her side. Inside the bowl was a damp, folded white linen cloth, the *chakin*, that she would use for purifying the tea bowl. A bamboo whisk (*chasen*) rested on the cloth, and across the rim of the bowl lay the tea scoop (*chashaku*). Bowing, Yuko said she would make tea for us. We placed our hands on the mat and bowed with her. She then carried the tea bowl to the stand and arranged it in front of the cold-water jar together with the caddy.

Yuko left the room once more and returned with a second tea bowl and the rinse-water container, the *kensui*. Inside this container was a ceramic rest for the kettle lid and, balanced on the rim of the *kensui*, a bamboo ladle, the *hishaku*. After crossing the threshold and closing the door, she knelt with

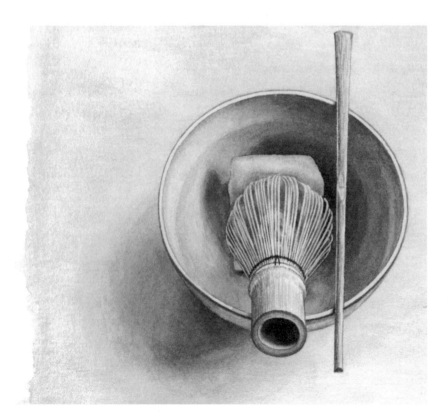

Figure 13

Tea bowl, cloth, whisk and scoop

these items and placed herself at an angle facing the hearth. She was following a defined sequence of movements and gestures that had been honed and handed down from one tea master to the next for centuries. Although the utensils were modern creations by local potters, her precise motions kept tradition alive.

Before purifying the tea caddy and scoop, Yuko took a deep breath to compose herself. In a tea ceremony, the way a utensil is handled represents the willingness and giving spirit with which the host serves the guests. Though all of the utensils were clean, wiping the caddy and bamboo scoop in front of her guests was an attempt on Yuko's part to purify her mind and heart.

As Yuko took out the red silk cloth from her *obi* and began to fold it, she established a rhythm with her movements. The ease with which she wiped the exterior of the caddy and set it down, then did the same with the tea scoop, induced a calmness in us. She took the bamboo whisk out of the bowl, set it aside on the tatami, put the lid of the kettle on its ceramic rest, and then placed the white linen cloth from the bowl onto the kettle's lid. Her movements became a flow of gestures that ultimately freed us all to share a level of rapport difficult to come by in ordinary life.

After pouring hot water into the tea bowl, Yuko carefully examined the tines of the bamboo whisk, then gently whisked the water. She placed the whisk on the tatami next to the caddy, discarded the water into the rinse-water container, and dried the tea bowl with the white cloth. It was time to put in the tea, but before doing this, she invited Jack to take a sweet. During the serving of thin tea, sweets are generally enjoyed as the host takes tea from the caddy and places it in the first guest's bowl. *Okashi*, traditional Japanese sweets, are typically served.

There are two kinds of *okashi*: dry sweets (*higashi*) similar to a crisp cookie, and *namagashi*, moist, soft cakes usually made of rice flour with a sweet bean paste filling. What makes these sweets unusual is their colors and shapes, which often resemble seasonal flowers or leaves. In general, sweets provide just the right contrast to the slightly bitter flavor of the tea.

Jack bowed to me and excused himself for "going first." After gently lifting the tray of *higashi* as a gesture of gratitude, he set it down and helped himself to one of the delicate wafers before moving the tray toward me. The stacked wafers, each with an imprint of pine and plum blossoms, had a subtle pink hue. While we passed the sweets tray and ate the crackers, Yuko put two scoops of tea from the caddy into Jack's bowl, then added just enough hot water with the bamboo ladle to fill a third of the bowl. She picked up the whisk and began to whip the tea into a smooth foam, creating a sound that complemented the song of the steaming kettle. Before setting the bowl of tea out for Jack, she turned it clockwise in two quarter-turns so that its "front" would face him.

This "fronting" of the bowl is very important, for in a tea ceremony the front should always face the host or guests as the bowl is passed from one to the other. I was once told that by respecting the front of the bowl, the host also respects the potter who made it. If a bowl has no apparent design that indicates the front, the host determines the front. This fronting is done as a bowl, tea caddy, or other utensil is passed not only between host and guest, but also from one guest to another. I once heard a student ask his teacher why it was important to front a utensil. The teacher was at first puzzled by the question. In Japan, it is taken for granted that the front of anything would face the person receiving it.

Having finished his sweet, Jack picked up the tea bowl Yuko had set out for him and placed it between him and me. He bowed and excused himself for being served before me. I bowed with him, silently inviting him to enjoy his tea. Moving the bowl in front of himself, he bowed to Yuko and responded in Japanese the equivalent of "I receive your tea." Jack placed the bowl in his left palm and, supporting the side with his right hand, lifted it in gratitude. Remembering to turn the front away from his lips—so as not to drink from that part of the bowl—he took a first sip, a second, and a third, and finished the tea with an audible sip of satisfaction. Before he set the bowl down, he turned it so that the front faced himself again. Pausing to observe the shape, design, and color, he then fronted the bowl and passed it to me, so that I could also enjoy the deep blue glaze that harmonized with the color of both the cold-water jar and the vase.

Examining a finely made bowl is one of the many pleasures of tea. One studies the shape; the contour of the lip or rim; the sides, or "hip"; and the bottom, or "foot," where the bowl touches the mat. Each aspect helps reveal a distinct whole. At the foot of a bowl, one might see the signature of the potter or glimpse the temper of the maker's hand. The best part of holding a bowl, I believe, is enjoying its feel, the warmth, spontaneity, and vitality a bowl conveys. When I set Jack's bowl down, I remembered that he had once said that he could "feel the age" of a utensil. I have never been able to so determine the age of a utensil, but when I set the bowl down it seemed subtly changed by my touch.

Meanwhile, Yuko used the second bowl to make my tea. It was white and deeper than the first one. When she had set it out on the mat for me, I moved forward on my knees and returned Jack's empty bowl to her in exchange for mine. Moving back to my place, I made the usual bows, joining Jack in taking tea, excusing myself to Anne, and thanking the host. As always, the aroma preceding the first taste exceeded my expectations. The smooth emerald froth against the white glaze of the bowl enhanced its pristine flavor and transcended the senses. Yuko smiled at me as she rinsed and dried Jack's bowl, then began to make tea for Anne.

Unlike the more formal serving of thick tea, where a certain quiet reigns, our casual thin-tea ceremony encouraged conversation. Although Jack always preferred silence, other tea students felt that silence could sometimes be as disconcerting as excessive chatter. As first guest, Jack had to be particularly sensitive about when to speak and when not to, and to establish a delicate balance of communication.

Everyday affairs are always left outside the tea room. When I had first met Jack, I was impressed by his knowledge of tea and ceremony and was curious to know the reason behind his dedication. He was a graphic artist in a fast-paced work environment, and to step out and remove himself from all that urgency was, he said, why he took to "tea and silence."

Finding an appropriate moment, Jack inquired about the calligrapher of "*jaku,*" tranquility, on the scroll, and commented on the camellia bud that seemed to have opened a bit more as the ceremony progressed. He wondered where the tea had come from. Yuko responded that it was from Uji. "And the first bowl?" Jack asked. "I purchased that one from a California potter who enjoys making tea utensils, even though he's an extravagant coffee drinker," Yuko said.

Anne followed the same sequence in exchanging my empty tea bowl for hers. During the conversation between Yuko and Jack, Anne sipped her tea and seemed introspective. I recalled seeing that same look in her blue-gray eyes many times. I had met Anne at a community college, where we were both taking a writing course. We had discovered tea at about the same time, and became fast friends because of our shared love of the traditional ceremony.

Anne finished her tea and returned the blue bowl. Yuko set it aside, then rinsed the white bowl in front of her with hot water. Jack ascertained that everyone was satisfied and asked her to close the ceremony. Yuko did not make tea for herself, but that was not unusual. In a tea ceremony, the host is trained to devote attention to the guests and to making and serving tea for them.

All that remained was for Yuko to cleanse the utensils and take them back into the preparation area. She ladled cold water from the ceramic jar into the bowl; after whipping the water to cleanse the whisk, she set the whisk down and discarded the water in the rinse-water container. The white folded cloth and whisk were returned to the bowl. Next, she cleansed the tea scoop with her red silk cloth folded in the traditional manner, before placing the scoop, turned over, on the lip of the bowl. Both caddy and bowl were arranged on the tatami mat in front of the cold-water jar, as before. When Yuko finally replaced the lid on the jar, Jack bowed and asked if we might *haiken* the caddy and tea scoop—a term used when a guest requests an opportunity to closely look at and handle the utensils.

"Of course," Yuko bowed. We watched as she refolded her cloth and purified the caddy again, then laid the scoop gently beside it on the mat to the left of the hearth, facing Jack. He brought the items to his place and, following the usual procedure, inspected them only after Yuko had left the room and closed her door. When it was my turn to examine the contemporary red-lacquered caddy, I was captivated by the way black streaks glowed through the lid. A respected teacher once wrote that some of the rarest and most cherished utensils were not necessarily the oldest or the most exquisitely detailed. What set a utensil apart might be its history or some unusual characteristic. Holding and examining the scoop in my hand brought back a desire to make my own utensils. Students of tea are always urged to do so. Creating a clay bowl or carving a tea scoop from a piece of local bamboo certainly increases appreciation and respect for the utensils made in past and recent times by potters and tea masters.

When Jack returned the tea caddy and scoop to their place by the hearth, Yuko rejoined us and said that the caddy was a present from a friend who had recently visited Japan, and that the scoop had been made by a former student who gave it the name of Spring Dusk. Jack bowed and thanked her for sharing her utensils with us, for the delicious tea, and for the memorable celebration. She bowed silently with him, picked up the caddy and scoop, and turned to face the lacquered stand. She positioned the caddy on the top shelf of the stand with the water ladle and lid rest and left the room with the tea scoop in her hand, crossing the threshold for the last time. Turning at the doorway to face us, she knelt, set the scoop down, and bowed deeply. We bowed with her. Then Yuko closed the door, and the ceremony was over.

The three of us stayed a few moments to enjoy the harmony of the room as evening light deepened the gold of the tatami mats. The floor felt warm and alive when I put my hands on it to bow to Jack before he stood to leave. Following him, I bowed to Anne, then stood and went to the alcove, where I knelt, bowed, and paused to reflect on the calligraphy—a rich black script against a white background, a symbol floating in infinite space. The camellia, which leaned toward the room, had not fully opened. It was prized because, once open, it lasted but a short while. After one final pause before the almost inaudible sound of the kettle, I slipped out of the small opening and back into my sandals, which Jack had thoughtfully placed on the stepping stone for me. A fresh breeze fanned the garden in the twilight. After Anne joined us, Yuko appeared and knelt just inside the small door to see us off, remaining there in silence until we were no longer in sight.

Figure 14

Kettle in hearth

Reluctant to break the spell, Jack and Anne returned slowly to the porch of Yuko's house to leave their sandals and retrieve their belongings. I followed even more slowly, feeling a deep calm. Lulled by the oncoming evening in the garden, we put away our white socks at the house and stepped into our shoes. As we walked toward our cars, my impressions of the present began to shift. What time was it? It was just before 6:00 p.m. Would the stores still be open? Yes. But for the moment, I still took advantage of the peace I felt with my friends before we went our separate ways.

Meanwhile, Yuko closed the tea house. She told me later that she had stayed after we left, sat in front of the kettle, and made herself a bowl of tea. She felt our presence even as she removed the scroll and flowers from the alcove, stored her utensils, and tidied the room.

Two

Practical Adaptations

5
Ways to Tranquility

Adaptations must spring from a very deep understanding, in order to avoid betraying the purity of the tradition or its power, or the timelessness of its truth.

Sogyal Rinpoche [1]

The ways to tranquility that I present in this chapter are adaptations of ceremony that offer meaningful methods of practicing tea and achieving a greater appreciation of ritual. Anyone can create an adaptation, but one needs to be sensitive to tradition, to the wisdom we find in Rikyu's four principles and seven rules. Influenced by this wisdom and by the various tea cultures of China, England, and Japan, these adaptations can be seen as first steps to the Way of Tea.

In adapting the tea ceremony, it is important to remember that, although the practice of tea embraces ethical or moral principles, it does not rely on religious beliefs. Though a tea ceremony, or a tea gathering, may produce an appreciation of spiritual mystery, the ritual is a way for the host and guests to simply share a bowl of tea and a meal in peaceful companionship.

[1] Sogyal Rinpoche, page 361

Any presentation of a tea ritual, no matter how easy it may seem, requires organization, commitment, and time. Just serving tea and sweets takes at least forty minutes, depending on the number of guests. More time and preparation are needed when serving a meal in conjunction with the tea. What makes all this worthwhile, and even essential, is the ultimate calm of the removal that tea provides along with the intimacy experienced. As we face a new millennium in which human conflict and time pressures seem to escalate along with the pressures of life, there is a need for removal, for this kind of leisure.

A tea gathering may be given at any time. Considering work schedules, it may only be feasible on a weekend or, in the summer, a late evening, perhaps on the night of a full moon. Gatherings may be spontaneous, but when planning ahead, as much thought should be given to invitations as to the choice of guests, food, and tea. Children may certainly be included, but if so, the child should be mature enough to appreciate ritual. I have discovered that most children are drawn to rituals, especially those that include sweets and different ways of drinking tea.

Implements and Location

There are two essential elements in making any style of tea: implements and location. Implements include utensils, the necessary furnishings, tea, food, and sweets to complement the tea's flavor.

Whether an adaptation is held in your home or in a tea house, it takes a certain amount of ingenuity to select the appropriate furnishings. Utensils are also a major consideration. They can be collected, purchased, or made by hand; or, with imagination and resourcefulness, you can discover tasteful substitutes. When choosing utensils, remember the wisdom of Rikyu: he insisted that it is not the costliness or antiquity of utensils that matters, but the attitude and dexterity with which they are used.

Certain implements are essential for an adaptation of whisked tea—mainly a kettle, a whisk, a tea scoop and tea bowl, and some good powdered green tea, enough to fill a displayed caddy, keeping in mind that each guest will need at least two scoops of tea powder. A waste-water container, a cloth for drying the tea bowl, and a fine silk cloth or napkin for the purpose of purification are also needed. Utensils for brewed tea are easier to find, and most people already have what is required: a kettle, a good teapot, and cups and saucers. According to Lu Yu of China's Tang Dynasty, the importance of good water cannot be ignored, so when making any kind of tea, it is best to use filtered or bottled water whenever possible.

An attractive container for either loose-leaf or powdered tea, the tea caddy is a requirement in both whisked-tea and brewed-tea adaptations. A caddy is especially needed when making whisked tea, and should be on the table when brewed tea is served so that guests may inspect the tea they are drinking if they wish to do so. It is as necessary to be familiar with the tea as it is to be sensitive to the needs and tastes of the guests. Such familiarity comes from the host's own discriminating taste in tea.

The consideration of location is essential to ceremony and tranquility. In tea, the place, the atmosphere, the role of host and guests, and the essence of the four principles all contribute to a successful adaptation.

Place

Supremely important is place, the area where the ceremony will occur, a place in which host and guest can be free from their individual constraints and enjoy together what they know about tea and ceremony. The experience of removal from our otherwise stressful lives begins with privacy. One of the primary attractions of the tea ceremony is that it offers a specific place for privacy, such as the tea house. However, tea houses are not readily available in the West. How then can one establish such a space?

One way that I have found to create removal outside of a definitive location is to remember that no matter where I am, I can take charge of my immediate space to some degree. Whether I am in a busy restaurant or alone in my own living room, I can occupy any space by becoming more aware of my place in it. At the same time, I may free myself from distraction by focusing attention on the table where I am seated. In this way, I can be less distracted by external noise or conversation. At home, one may simply close a window on outside traffic, or turn off an overhead light and use soft lamp-light for a more relaxed atmosphere.

Various locations may be suitable for adapting the tea ceremony. New practitioners may want to convert a spare room into a tea room at home. A porch, atrium, gazebo, or patio might be appropriate. If these are not available, an area within a large room can be used for tea gatherings. If climate and economic resources allow, perhaps a simple tea house could be built outdoors. But in doing so, carefully consider basic Japanese design and the use of natural materials, together with a sense of seclusion. The tatami mats used by the Japanese are very appealing because of their natural texture and scent. Each mat is approximately three by six feet in size; the smallest tea room may use two mats, allowing space for the host to make tea and seat at least two guests. Whatever arrangements you choose, you will want to ensure a flow of ceremony. Of course, this means unplugging the telephone or at least turning down the volume of its ring to ensure against sudden interruption.

Certain aspects of the tea ceremony should always be honored. This is where the practical rules of Rikyu are so valuable. When preparing your place, remember Rikyu's fifth and sixth rules: *do everything ahead of time* and *prepare for rain*. Above all, have regard for your guests by always allowing for the unexpected. For example, should an outdoor tea be organized, an alternative place should be kept in mind in case of rain.

Atmosphere

In summer suggest coolness, in winter, warmth. Rikyu's fourth rule was not written with modern heaters or air conditioners in mind. Rather, he was suggesting that if some element of the tea ceremony hints at warmth in winter and coolness in summer, the experience as a whole will be heightened. In an adaptation, when the use of a stone basin of water is not feasible, you may want to offer your guests hot, damp towels in winter or cool, slightly wet refreshing cloths in summer. It has been said that the best way to follow this rule is to take to heart the suggestion of a host who tells the guests to "please be comfortable" at the beginning of a traditional ceremony. For the host, this is more of a compassionate wish than mere etiquette.

No matter where you serve tea, carefully consider the lighting. Use daylight whenever possible. Indirect lighting is always best, and not too much—just enough so the effect is subtle, more intimate. Afternoon or evening candles provide a certain magic and warmth. Wavering candlelight casts a glow that enriches objects and justifies either silence or soft, appropriate music. Candlelight often reminds me of Jun'ichiro Tanizaki's observation that "were it not for shadows, there would be no beauty." [2]

Although music is seldom part of a traditional tea gathering, flute music or a fine Segovia album can help define the dimensions of a room, as well as calm and please your guests. However, because we always want to remain sensitive to the ceremony's most important sounds—the steaming kettle and the preparation of an individual bowl or cup of tea—you may find it best to turn the music off just before the ceremony begins.

In winter, using a narrow flower vase arranged with a seasonal flower adds an intimation of warmth. In summer, a wide basket of different grasses and blossoms offers a cool feeling. Flowers are an offering, a way of welcoming guests to the room. Rikyu reminds us to display flowers by placing them in a container as they might look in the field. One of my teachers once

[2] Tanizaki, page 30

said that when arranging flowers for tea, "you use your feet." In other words, you walk into a garden or field and select seasonal blossoms that are not distracting, ostentatious, or scented. Once collected choose from among the flowers, arranging them in your hands as they would naturally appear. Then with one movement place them forward facing in the vase, caringly, welcomingly. As words on a scroll teach us human wisdom, we learn from flowers the wisdom of nature. The consistency of change, as in the seasons, and the dew that appears remind us of their ephemeral life.

The Host

Anticipating and fulfilling the needs of the guests is the primary responsibility of the host. In an adaptation, as in a traditional ceremony, a mindful host creates an atmosphere in which the tradition and ritual of tea can be maintained.

As the host, choose a theme for the ceremony and try to reflect that theme in a painting, a piece of calligraphy, or any appropriate artwork you may have at hand. An attitude of consideration, temperance, humility, and simplicity instills greater harmony and grace into the occasion. Following the first rule—to *make a delicious bowl of tea*—depends on a sensitivity to the guests' needs, which also includes an awareness of problems with language, custom, or other differences among the guests. Ultimately, the efforts of any host are never a burden, but a matter of meticulous attention, especially to thoughtful spoken or unspoken communication.

You may include any number of guests in the tea gathering, but I have learned that it is best to begin by inviting only one or two people. Where guests are greeted and how they are seated may require some innovation and imagination, especially with a ticklish problem such as how to ensure the proper height of a table or chair. Finding a suitable tea strainer can be as much of a dilemma as selecting the appropriate sweets or sandwiches.

When the guests arrive, the wise host will want to recall Rikyu's seventh rule: *Give those with whom you find yourself every consideration.* Your role as a host is never alone in this effort, for discerning guests at a tea ceremony always give the same consideration to each other.

The Guest

Since a pleasant outcome to any tea gathering depends largely on the seamless interaction of guest and host, many teachers say that the role of one is that of the other: the roles are equal. This kind of intimate exchange is fundamental to the tea ceremony. In an adaptation, the guest needs to be sensitive to tradition, rules, etiquette, and each effort of the host, as well as cognizant of the freedom or limitation that an innovative approach can provide. Every tea gathering—even tea for two—requires a host and guest. Ultimately, it is about relationship, intimacy, and making that heart-to-heart connection that dissolves separation. Thus, tea brings us closer, opening a way to intimacy and sensitive consideration of others.

Practiced guests usually arrive five or ten minutes early. After a gracious greeting and bow, the host shows them where to leave coats and purses and, if necessary, their shoes. Traditionally, a basin of water and a ladle for cleansing are set out in an appropriate place outdoors, so that guests are able to rinse their hands away from the basin. This can be done indoors in a washroom if the tea gathering takes place in an apartment. The act of purification, of rinsing the hands and mouth, provides a moment alone before approaching or entering the area that the host has prepared for tea. After taking a little time to compose the mind and let go of the busy cares of the day, guests become more conscious of sight and sound as they approach an arrangement of fresh flowers or a scroll that embodies the host's theme. If no other guest is waiting, this scroll deserves an extra moment of appreciation. Before sitting down, guests may stop to admire a beautiful utensil or the tasteful setting in which the host has displayed a teapot or kettle.

Guests may be seated in a narrow studio apartment, in the corner of a large, windowed living room, or outside on a quiet deck. Seating for a gathering can also be arranged around a small coffee table or in a larger dining room. Whatever the arrangement, guests enter into the spirit of the place as well as the ceremony. At the end, before leaving the area or the room, they are encouraged to take another moment to appreciate the atmosphere that the host has created.

The Essence of the Four Principles

As you incorporate Rikyu's principles of harmony, respect, purity, and tranquility into a modern adaptation of ceremony, you should also envision Rikyu's seven basic rules as steps along the way.

In the tea ceremony, *harmony* is the warm congeniality that arises from the intent of both host and guest to create a peaceful removal. Harmony is also experienced in the theme, the selection of all the necessary items, and the spirit in which tea is made, served, and received.

Respect, in traditional Japanese tea gatherings, is paramount. Bowing, the most pronounced form of respect, is second nature to the Japanese; it is a way of expressing deference and gratitude, of giving thanks to all things—to the earth, sun and moon as well as to tea itself. In other parts of Asia—India, for instance—it is common practice to bring the palms together and bow to another person, saluting the divine within that human being. Respect encourages all to see the interdependence of one another and the earth upon which they live.

In the West we shake hands, a form of respect that offers the advantage of touch and also conveys warmth. In an adaptation of tea ceremony, we can greet one another with either a bow or a handshake, expressing appreciation in ways that remind us of the interconnectedness of everything. When we receive a bowl or cup of tea, we can pause for a moment to thank the host and, before drinking, follow tradition and raise our cup or bowl in gratitude to all that made the moment possible. Such small acts are not only integral to the experience, they are also ways in which we appreciate the old and the new.

Purity, the third principle, represents cleanliness and a distinct state of mind. A host preparing the place and the utensils for tea concentrates on the task with the same attention as when wiping a tea caddy with a folded cloth in front of the guests. Although the caddy might be immaculate, this act of purification frees the mind from self-concern and confusion. Clarity arises, and with it, immense gratitude and generosity. The guests too, experience this widening of heart and mind and with it deep sensitivity, especially to sight and sound. Listening to the sound of water being poured, we are reminded by Dr. Sen, the fifteenth-generation Urasenke tea master, that in this pure sound is "the realm of nonattachment," and "to enter this realm is one reason why we practice over and over again the same procedures in making tea. . . . Making and drinking a bowl of tea involves no right and no wrong. It is a simple, open, and honest meeting of minds, beyond wisdom, experience, and point of view." [3]

With harmony, respect, and purity, *tranquility* comes naturally. All of the principles imbue ceremony with the purpose of setting aside the material concerns of everyday life; when all four are unified the experience becomes memorable.

A Touch of Patience

With regard to the two kinds of tea—loose-leaf and powdered— I separate the following tea adaptations into brewed and whisked. Brewed-tea versions include Chinese and English ways of taking tea, whereas the whisked-tea adaptation I describe is a table-style innovation of a simple Japanese ceremony (*ryakubon*). Although brewed and whisked adaptations differ from one another, both respect the enchantment that the beverage and setting elicits.

[3] Soshitsu Sen XV, *Tea Life, Tea Mind*, page 58

Depending on the time of day, season, or occasion, a meal or light snack can be served in conjunction with any of the adaptations. For instance, brewed tea can complement lunch, dinner, or an afternoon variety of finger sandwiches and small cakes. Whisked tea can be offered after a *kaiseki* meal of soup, rice, and seasonal vegetables. The focus, of course, remains on the ritual and the removal it offers.

6
Brewed Tea

Brewed tea differs from whisked tea in two fundamental ways: convenience and variety. Brewed tea is easier to prepare, and a greater selection is available. Brewing a pot of tea for several people is much easier than making individual bowls of whisked tea (although, as we have seen, there is more to serious consumption than the choice of tea or the method of its preparation). The hot, liquid transparency of a brewed cup offers a quality of moment quite different from that of the warm jade froth of a whisked bowl.

Brewed tea is, of course, the most popular kind in the Western world. Good powdered green tea is rather difficult to obtain in the West, and the better grades can be costly. Even in contemporary Japan, where powdered green tea is more available, the vast majority of people drink brewed tea and sometimes indulge in *sencha*, a brewed-tea ceremony. This is a change, for it was not long ago that small stone grinders used for turning green tea leaves into powder could be found in many Japanese homes.

Loose-leaf tea can be purchased in most supermarkets, specialty shops, and natural food stores in the United States. However, selection and quality vary. Determining quality is a matter of taste and study. If you are exploring the world of brewed tea, I suggest using loose-leaf as much as possible, since it is usually better and more economical than bagged tea. However, if a favorite blend is available only in tea bags, you can trim the string off for brewing or remove the tea from the bag to display it in a caddy.

One way to explore the full flavor of a tea is to brew it first strong, then weaker, and drink both versions without milk or sugar to taste the true flavor. When planning a ceremony, you will want to be aware of those guests who are sensitive to caffeine, drink only herbal tea, or prefer to add cream, sugar, or lemon.

Two adaptations of brewed tea follow: one of the Chinese *gong fu* ceremony, the other an adaptation of English tea.

Chinese Tea

> *The art of tea is artless in that it is practiced with the maximum of informality and freedom from restriction.*
>
> John Blofeld [1]

A Taoist friend of mine in San Francisco once pointed out to me that one of the differences between the Japanese and Chinese appreciation of tea is that the Chinese are more interested in tea's pharmacological or medicinal properties than in the spiritual implications of ceremony. My well-read friend believed this was due to the influence of Lu Yu. To be sure, Lu Yu was concerned with the way tea was prepared and with the selection of proper utensils. As we saw in Chapter Two, he put great emphasis on the quality of the water and the care with which it was boiled. He created a formality, an order for drinking tea, and thus a tradition. My friend saw the formality as more of a performance or an etiquette, which perhaps inspired an art rather than a Way of Tea. He said that the Way never became fixed in China, as it had in Japan, because throughout its history Chinese culture had been fractured—first by unending struggles among China's regions and dynasties and later by the interference of European trade and Western politics, which

[1] Blofeld, page 143

continue into the twenty-first century. As a consequence, a definitive Way of Tea, such as *Chado* in Japan, did not take hold in China. However, because of Lu Yu, the Chinese are still foremost among those who appreciate tea. Today, what is known as *gong fu* tea is still a popular practice.

Gong fu, or *kung fu*, denotes any "activity requiring time and effort to achieve mastery." [2] The tea ceremony was developed in the middle of the Ming dynasty (1368–1644), when tea drinkers began using shallow bowls and the Yixing teapot became fashionable. *Gong fu* is considered an art or activity of the literati, scholars, and connoisseurs of tea.

There are many reasons that tea is still considered an art in China. One is the influence of the poetry of the Tang and Song dynasties, which allude to misty mountains, charming gardens, and the beauty of the ceramic forms created especially for tea ritual, as well as the invigorating effects of the tea itself. Adapted and practiced in Hong Kong and Taiwan, as well as Malaysia, Singapore, and Thailand, *gong fu* ceremony is less formal than *chanoyu*. It is a way of brewing a fine green tea, or an oolong such as *Tikuanyin* from China or Taiwan, in a relaxed, congenial atmosphere. *Tikuanyin* oolong has a memorable flavor and aroma, especially when prepared in small pots like the Yixing and served in tiny cups that are not much larger than those in which the Japanese serve sake.

A *gong fu* ceremony is simple yet graceful. Venturing to create an adaptation of this ceremony is rewarding, and assembling the right utensils is fairly easy. When preparing for your guests, however, it will be helpful to refer to the illustration. What will be needed is

> ☙ **A table with enough space for the host to manage all the utensils efficiently.** A variety of specially designed and beautifully crafted tables are available in Western countries. They may be purchased in certain Chinese stores that carry fine tea or furniture.

[2] Blofeld, page 158

Figure 15

Gong fu Chinese ceremony

⟋ **A pot of hot water on a hot plate.** As shown in the illustration, this is placed on the table to the left of the host. The pot should be an attractive one; however, if this is not feasible, a flask of very hot water can be substituted.

⟋ **A large, deep pan with a drain to accommodate waste-water.** This is built into the table, directly in front of the host. (A chafing dish with a perforated insert as a waste-water pan will work just as well.)

⟋ **A tea-boat, a shallow bowl that holds the teapot.** This "boat" sits in the pan and is the focal point of the table.

🍃 **A small teapot that can fit inside the tea-boat.**

🍃 **A tray of cups and saucers.** These are laid to the right along with other important implements and utensils.

🍃 **A rolled tea towel** is placed behind the china.

🍃 **A smaller tray with a tea caddy, a spoon, and a dish for the tea leaves** lies on the end of the table.

For my adaptation, I used a sturdy round dining table and tried to mirror the arrangement in the illustration. I began by putting a small kettle of water on a portable burner to my left. To my front I placed a long chafing dish for my waste-water pan. A shallow bowl was placed in its center. Inside of this bowl or "tea-boat" I set a miniature Yixing teapot I found in San Francisco's Chinatown. I then arranged a small tray with a set of four delicate cups without handles, and matching saucers. These were a Christmas present from my friend Jack. (I never learned where he had found this prized set of white china, as he rarely divulged such secrets.) Behind these cups, I set a tea towel and, to the right, another smaller tray. On this, I put a spoon and a dish for displaying the tea.

To begin my preparations, I filled a comely stainless steel caddy with *Tikuanyin*, an oolong known for its digestion-aiding properties. I then prepared some damp, hot towels and put them on a bamboo tray. These would be passed to my guests before starting to make the tea.

The invitations I sent were handwritten and included a short, simple poem hinting the theme of the gathering:

> *Immortals*
> *do not measure*
> *life with time*
> *but with breath and season.*

Tea and...

The invitations gave the date, time, place, and names of the guests. The guests included a married couple, David and Mary, who enjoyed Chinese tea and had expressed much interest in the *gong fu* ceremony. Jack was also included because he had introduced us. The three guests—Jack, Mary, and David—came over after a brunch celebrating Mary's birthday. She had turned fifty, so I displayed a scroll I had bought in Hong Kong many years before: a painting of a beautiful crane on a pine branch representing longevity. Beside the scroll on a side table was a miniature bonzai, intended as a gift for Mary, who enjoyed cultivating bonzai.

When they arrived and sat down, I brought in the damp towels and placed them in front of Mary. She smiled, took one, and passed the tray. When the towels were used I removed the tray and brought in a dish of sweet cakes.

After a moment of silence, I took the four cups off the large tray and lined them up on the perforated insert of the chafing dish next to the shallow bowl. Picking up the kettle of hot water, I "scalded" and purified the cups and teapot by filling them to the brim. My Taoist friend believed that "the hotter the pot, the better the brew." As the pot and the cups warmed, I spooned some tea from the caddy into the small dish, part of the customary procedure of inspecting the leaves for impurities. Jack had taught me how to examine tea leaves. Knowing how much he enjoyed this part of the ceremony, I offered him the dish of *Tikuanyin* for inspection.

He looked at the leaves, praised their scent, and said that the tea seemed perfect. Smiling, he told us that *Tikuanyin* means "Goddess of Mercy." Kuan Yin, he explained, was known in Japan as Kanzeon, "the listener of human cries," and in India as Avalokitesvara, the many-armed deity who reached out to help those who suffered.

When I emptied the teapot of hot water, Jack handed back the dish of leaves. With a new understanding of this oolong, I slid the leaves into the pot and again added more hot water.

The Chinese pay strict attention to the correct water temperature, which varies depending on which tea is used. Tea water must be as hot for oolong as for green tea—just as it begins to boil (when the temperature reaches about 160° to 180°F). The right water temperature allows as much flavor as possible to come from the leaves.

When a green tea is chosen, use enough to fill a third of the pot. For oolong tea, the leaves should fill approximately half of the pot, and the hot water should be poured carefully so that it falls evenly over the leaves.

The first infusion, or soaking, is to "wash" the leaves, and this water is immediately discarded into the waste-water pan. Of course, there should be a strainer built into the tea pot to hold in the leaves. More hot water should then be added, the lid replaced, and the leaves allowed to brew for a maximum of thirty to sixty seconds. Green tea requires a few seconds longer to brew than oolong. Care is taken that the brew does not get too strong. The use of a watch discreetly placed on the table to keep track of the seconds is helpful. Pour more water over the outside of the teapot to keep the tea hot, until the "boat" or bowl in which it sits is half full.

As I discarded the first infusion which soaked the leaves, I then added more hot water and recalled the words of an English Zen monk who once exclaimed of the leaves, "Oh, bathe the beauties, this is their moment of glory." While the leaves brewed, I scalded the cups once more to keep them warm. Then I tried something I'd seen a Chinese friend do. He gingerly rolled each cup lightly against the handle of the teapot while emptying the cup into the "boat," before placing it on a saucer. This produced some interesting tones. Once the tea was poured into the cups and served, the remaining brew was discarded. We sipped what could easily have been finished in one gulp, relishing each sip in a way that Lu Yu suggested long ago, as if it were life itself.

Also on the advice of my Chinese friend, I took care to steep the subsequent brews for less than fifteen seconds, to prevent them from becoming bitter. The same leaves gave us each four cups of glowing amber tea.

The *gong fu* is a relatively easy ceremony for anyone to adapt, and finding the appropriate utensils can be half the fun. Charming Yixing teapots, or something similar, along with a set of diminutive cups, are not hard to locate, but any small teapot and cups will do just as well. This is always a delightful ceremony to share with friends.

Toasted pumpkin or sunflower seeds, roasted nuts, or small cakes on individual dishes are sometimes served before the tea is made. For Mary's birthday tea, I served mooncakes—small cakes stuffed with sweet bean paste or lotus root—purchased in an Asian store.

An Adaptation of English Tea

In the past two hundred years, the ritual of English tea has spread throughout the world. It has proved especially popular in the Far East. However, the etiquette of English tea created by Anna, Duchess of Bedford, has been modified to accommodate the limits on place and time that people have to devote to "afternoon tea."

English tea, like the Chinese *gong fu* ceremony, has its own graceful and elegant aspects that can help guests achieve a unique removal. In the colonial era, "having tea," whether in jungles or in drawing rooms, never failed to confirm the "Englishness" of countless women and men who, as they drank, did not know whether they would ever see England again. Today, English tea is familiar to many people in the West and East and has become a daily necessity in the lives of millions of people.

"Taking tea" has long had an important social purpose for the British. At tea, people can come and go, laugh and eat. Two guests might go off to a corner for a private chat, while another may drop in just to make an appearance, greet the host, politely have a cup, and then leave. In looking for a new way to serve English tea, we should bear in mind that the tea is generally not made in front of the guests, but prepared ahead of time in the host's kitchen. When the guests arrive, the tea is often ready to pour. Convenient as this is, it is obvious that some ceremony and an appreciation of the actual *making* of the tea are missing.

Considering this, I wanted to create a different spirit of ceremony. If I made tea in front of the guests, I would need a source of hot water, such as a kettle on a hot plate placed on or near the main table, and a teapot. When cups, saucers, cream jug, and sugar bowl are used, along with plates, forks, and spoons, the table can become too cluttered for the graceful execution of a ceremony. For this reason, I felt the need for a basic variation that would include some elements of an Eastern tea ritual. In a phone conversation the day before my English tea adaptation took place, Jack remarked that this said more about the purpose for a removal than it did about English tea.

Rather than having a social tea, the significance of adapting the English ceremony is to produce a basic yet distinct removal, using the fine black leaves served at an English tea. It occurred to me to offer this adaptation one day when my friends Christine, Hugo, and Ellen planned to stop by before going to their monthly poetry meeting. I thought the adaptation would be a nice prelude to their meeting. A week before the meeting I called and invited them to what I explained would be an English ceremony of brewed tea. They were interested and happy to participate. After I invited them, I sat down and made a list of what might be needed—utensils, furnishings, food:

  ☙ A source of hot water, perhaps a portable burner.

🕉 A kettle and teapot, preferably one with a built-in strainer, not too garish or impractical but with some character.

🕉 A fine wire-mesh strainer.

🕉 A tea caddy.

🕉 Tea, a bowl of sugar and a creamer—simple and small.

🕉 A display of flowers and/or a piece of calligraphy.

🕉 A table suitable for tea, perhaps a wooden card table.

🕉 Chairs.

🕉 A towel or folded cloth for pouring tea.

🕉 A wooden or lacquered tray for utensils.

🕉 Cups, saucers, and teaspoons.

🕉 Small plates or a platter for edibles—simple sweets, cheese, cucumber sandwiches, fruit, or plain cake.

I was concerned about my furnishings and artifacts until I recalled Rikyu's belief that tea was "simplicity," and that what was important was to "use just what utensils you have." To eliminate the problem of hot water at the table, I decided to brew the tea in the kitchen in a white china teapot and bring it in as it was steeping. This teapot had a strainer inside at the base of its spout, and a candle warmer in its ceramic base. I would initially display the teapot on the table with just hot water in it to warm the teapot.

Although I would not make tea at the table, I still wanted to display a caddy: a round brass tin that had a black-lacquered lid with a red leaf-and-flower motif. I had originally thought of serving Glenburn Darjeeling, packaged by the Republic of Tea Company, but switched to Keemun, a Chinese black tea, that Hugo had sent over the day before as a gift for the occasion. Most people prefer Keemun without sugar or cream, so that eliminated the need for a creamer and sugar bowl.

I chose "freedom" as a basic theme—freedom from constraint, yet the freedom to use traditional aspects of the Way of Tea that had become so meaningful. I had initially planned to display a scroll with the hand-printed phrase

> *Distant clouds come and go*
> *But the blue mountain*
> *remains motionless.*

I loved this Zen saying, but since this was English tea, I decided on a Virginia Woolf quote:

> *To enjoy freedom,*
> *we have to control ourselves.*

Not only was Woolf English, but her expression fit the theme of the gathering better. I copied her quote in black ink onto an oval of gold-bordered pasteboard and propped it on a small wooden easel. To accent it, I picked delphiniums and pink cosmos from the backyard and arranged them near the easel in a slim, unglazed vase on a side table against a wall of the living room.

Just before my guests arrived, I covered a round table near the window with an ivory lace cloth in order to soften the afternoon light, and arranged some chairs.

๑ I lit a candle in the warmer and placed the teapot filled with hot water on top of it.

๑ The teapot was centered on the table.

๑ At my place I set the tea caddy, a folded cloth, and a tray with four ceramic cups.

In the kitchen, I arranged a dish of pitted dates, dried apricots, and slices of preserved papaya on a light green stoneware dish. I had thought about cake, but both Christine and Hugo loved fruit. I also prepared a tray of damp, cool hand towels to offer them once they were seated.

Figure 16

An adaptation of English tea

We were having a mild summer, but the sky was filled with clouds and the air smelled of rain. In case it rained, I put a tall ceramic holder for umbrellas by the door just as Christine, Hugo, and Ellen arrived. Taking their coats, I showed them where to sit and went to fetch the cool towels, giving them time to relax. As I did, the three of them paused before the Virginia Woolf quote and vase of bright flowers.

Once I knew they were seated, I brought in the tray of towels, set it down before Hugo, then went back for the dish of fruit. Hugo, who had some knowledge of Japanese tea etiquette, thoughtfully bowed to Christine, excusing himself for going first. Only then did he pick up the tray of towels, raise it slightly, and set it down. Taking a towel, he passed the tray to Christine and Ellen. When I returned to the table with the fruit dish, Hugo looked at the dish and smiled. I invited them to help themselves, removed the towels, then picked up the teapot and returned to the kitchen. In the kitchen

๑ I emptied the "first water" that warmed the teapot into the kitchen sink,

๑ then carefully measured some of the Keemun into the pot, and added more hot water.

๑ As the tea steeped, I took the pot to the table and joined my guests.

๑ I set the pot on its ceramic warmer and moved the tray of cups to my left.

Hugo helped himself to a slice of papaya and passed the fruit to Christine. Lifting it in thanks as he had done, she set the dish down and took some apricots before passing the dish to Ellen, who selected a date and then turned the dish toward me. I enjoyed the contrast between the green glaze of the dish and the gold papaya. Taking a slice of it, I handed the dish back to Hugo.

Preparing to pour the tea, I moved the tray of cups in front of me. As I filled them, the sound of the liquid seemed to enhance the quality of an agreeable silence. We each took a cup, bowed, and raised them in appreciation of the tea and the moment together before sipping the brew. The texture of smooth brown glaze felt warm and comfortable. When Christine asked about them, I told her a local potter had made the cups. Hugo said that they added to the effect of the tea, which was not strong but had a "special character" that insisted on "being slowly savored." As we all drank a second cup of this Keemun, a steady rain began to fall. The trickling sounds reminded me of a simmering kettle in a Japanese tea house.

Hugo asked whether he might have a closer look at the tea caddy. I told him it was a favorite of mine, and I had wanted to display his Keemun in it. He was amused when he peeked at the underside and found that it came from a thrift shop in Ashland, Oregon. Opening it, he sniffed the tea and then told us we were drinking sweet Imperial Keemun from the mountains of Anhui. He had purchased the tea from a shop called the Tao of Tea in Southeast Portland. The owner considered this "a good tea," but admitted it was not "the greatest." According to him, the finest Chinese tea is all but impossible to come by because it's grown in "sacred gardens in ridiculously small quantities." Apparently the best tea was "closely guarded," and was available only to "important Chinese officials and a few well-connected connoisseurs around the world."

When we finished, a leisurely hour later, I cleared the table, leaving the teapot and caddy as centerpieces. Returning from the kitchen, I thanked my friends for joining me and blew out the candle in the warmer before leaving the table. They lingered for a few moments, then stood before the Woolf quote and flowers, bowed, and left the room.

7
Whisked Tea

One of the reasons the Western world never became familiar with whisked tea is that Europe began to use tea only after the popularity of powdered tea had waned in China. [1] The bamboo whisk was used in China during the Song dynasty (A.D. 960–1280), when unfermented green tea leaves were ground into fine powder. Okakura mentions that poets of the southern dynasties called a bowl of whisked tea the "froth of the liquid jade." Cake tea was boiled, powdered tea was whipped, and leaf tea was steeped, practices that, Okakura said, reflected "the distinct emotional impulses of the Tang, Sung, and Ming dynasties of China" and also might be described as "the Classic, the Romantic, and the Naturalistic schools of Tea," respectively. [2] After the Mongol invasions, manners and customs in China changed so much that by the close of the Ming dynasty in 1644 powdered tea and the whisk were all but forgotten.

The whisk fared better in Japan. Ever since powdered tea was introduced by Zen monks returning from China, the whisk has been used in the Japanese tea ceremony. The intrepid monks and other travelers interested in cultural exchange returned with an array of Chinese ceramics. Those from

[1] Okakuro, page 30
[2] Ibid., pages 21, 22

the Song dynasty are especially treasured as some of the finest examples of Chinese art.

Exquisitely crafted, the whisk, or *chasen*, is the best utensil for blending powdered green tea in hot water. Indeed, its design is so ideal that it has remained essentially unchanged for hundreds of years. Depending on the bowls used for making tea, the height of the whisk may vary from four and a half inches to six inches or slightly more. The whisk is handmade out of bamboo, a material valued for its resilience, lightness, and lack of scent. Bamboo, which is a kind of grass that grows abundantly in Japan on hills and mountainsides, has long been a part of Japanese culture and is used in construction, utensils, and for decorative purposes. Some of the finest whisks are made of bamboo that is over two centuries old, but these are quite rare. Whisks of ordinary bamboo are more available, but even they are individually crafted.

Traditional Japanese Table-Style Tea

At an international exhibition in Kyoto a century ago, members of the tea community recognized the necessity of adapting the tea ceremony to accommodate Westerners. It was a time of transition from the Edo to the Meiji period in Japan (1868–1912) when all things Western were heavily promoted. Gengensai (1810–1877), the Eleventh Generation Grand Tea Master of Urasenke, who formally opened the Way of Tea to women—a way that had once been the domain of the samurai—introduced a ceremony using tables and stools for the host and guests. [3] Later, the grandson of Gengensai, Tantansai Soshitsu, went a step further when he revised *ryurei*, a form of traditional table-style tea that is practiced today. This was a remarkable change from traditional Japanese tea, which was served on the tatami floor of the tea room.

3 Soshitsu Sen XV, "The Diffusion of the Way of Tea," page 6.

Table-style tea is not generally offered in a traditional tea room but, out of regard for the fragile tatami mat, is usually conducted out of doors, in a pavilion or a room with a different floor covering. Most of the procedures remain the same: the silk cloth is folded, the utensils are purified, sweets are offered, and the powdered tea is whisked. A kettle sits either on a table to the left of the host or on a round metal built-in recess where it can be heated by a charcoal fire or a portable electric burner.

One type of table used is the *misono dana*, or "imperial garden stand," from Japan. It is black-lacquered and trimmed with red cord. Built into the table (on the left side facing the host) is a round hollow recess on which the kettle sits. The recess holds a copper basin for a charcoal fire or an electric appliance. Light and portable, the *misono dana* comes in three parts for assembly in or out of doors. It also has a shelf on which the host can store a rinse-water container and extra tea bowls. When it is used out of doors, a huge red umbrella is usually set up beside it. Long tables, lower than usual, are covered with red cloths and serve as benches for the guests. These offer comfortable seating and a place for the guests to put their bowl of tea or plate of sweets. The outdoor setting allows pleasant and colorful gatherings, especially when it is complemented by kimonos worn by hosts and their guests.

Table-style tea requires as great a commitment to study and practice as does traditional ceremony. While students master the procedures, they also learn about the history of the tea tradition and its philosophy. These important facets of the Way of Tea require the counsel of a trained teacher in order to ensure that complete attention is paid to such things as the proper use of the wiping cloth and the adept handling of utensils. Until a teacher is found, it is probably wiser, and easier, to adapt a form of table style whisked tea—as long as there is an appreciation of basic principles, as well as a whisk and some powdered green tea.

An Adaptation of Japanese Table-Style Tea

When I became interested in the adaptation of a Japanese table-style tea, I wondered whether powdered green tea could be used in the West without some inquiry into the Way of Tea. Anyone can learn how to whisk a bowl of tea, but to the beginner the simplest ceremony may seem intricate, especially in the folding of the silk napkin and purification of utensils. In order to achieve a measure of ease and confidence when presenting this tea, the basic purification procedures require practice and patience. It is best to begin by learning the way of purification before going on with the actual making of tea. Here, the guiding hand of a teacher is invaluable. [4]

For most people, the taste of whisked tea is motivation enough to seek a shared moment of peace. In Japan, powdered tea, *matcha*, is processed mainly for tea ceremony, and for this reason is not as widely distributed as loose-leaf tea. Some select stores in the United States and other Western countries do carry good powdered tea and accept mail orders. [5]

Putting together a whisk, a scoop, a bowl, and some tea is hardly the secret of a dedicated few. But enjoying whisked tea requires anyone, Westerner or Easterner, to learn its proper preparation. When whisked tea is sensitively shared in a ceremony, it produces a removal, a space of rare peace. With this in mind, I planned a simple version of a Japanese table-style tea, one that retained traditional procedures and that a person unfamiliar with whisked tea could adapt with practice.

First of all, I decided to invite guests who were inspired by ceremony and new ways of drinking tea. The next consideration was whether to make tea in or out of doors. This would depend on the weather, of course, and on whether an available table would provide space for the guests and the making of tea. If the table was too small to accommodate everyone, a tray-table could be set before each guest for serving the tea and sweets.

[4] See Some Sources
[5] See Some Sources

To accommodate a host without formal training, I have modified certain customs. The traditional folding of the wiping cloth used to purify the caddy and tea scoop, as well as the *hishaku*—the bamboo water ladle for drawing hot water from the kettle—are difficult to master without some study and practice. For this adaptation, there is no need to use the *hishaku*, but traditional purification procedures should be observed.

There are a few basic steps to remember in this adaptation:

⮑ Decide whom to invite, where to have the tea, and when to have it.

⮑ Set up the tea table and arrange the room.

⮑ Assemble a serving tray with utensils in the kitchen.

⮑ Make and serve sweets and tea to the guests.

⮑ Close the ceremony. After the guests leave, clean up and put away utensils.

The essential elements for a presentation of table-style whisked tea are

⮑ A tea table.

⮑ A low, small stand by the table for a rinse-water container.

⮑ Stools, chairs, or a low bench; around a coffee table, cushions might be used.

⮑ A source of heat for the water—ideally an iron or ceramic brazier, or any receptacle that holds hot coals or a single burner (whatever is used should be placed on a heat proof stand or surface); or a thermos can be used, preferably one that allows for hot or near-boiling water.

◑ An iron kettle to heat water (any kettle will do so long as it lets steam escape but does not whistle).

◑ Powdered green tea.

◑ A fine wire mesh screen to sift the tea.

◑ A tea scoop.

◑ A presentable tea caddy for the powdered tea.

◑ Tea bowls or small bowls in which a whisk can be used.

◑ A slightly damp tea cloth, or *chakin*, for wiping the tea bowl.

◑ A bamboo whisk.

◑ A round tray for the utensils (about eleven inches or so in diameter).

◑ A square cloth, or *fukusa*, for purifying utensils (preferably silk, with a width and length of about ten and one-half inches).

◑ A rinse water container (a ceramic or brass receptacle).

◑ A plate, tray, or ceramic dish for edibles.

◑ Edibles: sweets or fruit.

◑ Small folded paper napkins.

◑ An appropriate wall area or alcove for a scroll, painting, photograph, a vase of flowers, or a combination thereof.

◑ A large basin of water, or *tsukubai*, with a ladle for guests' use.

I decided to invite Jack and Anne, who by then were as intrigued with the idea of adaptations as they were by some of the decisions and concerns I had in planning the ceremony. Jack asked if he might bring a friend who had expressed an interest in tea ceremony. "Patricia," he said, "loves tea." I was delighted to include her as well.

When the day arrived, it was cool outside and threatened to rain, so I gave up the idea of an outdoor gathering. I then had to decide where to put the table, for I required not only a nearby outlet for a single burner, but a setting that offered some kind of ambience.

Traditionally, charcoal is used in a brazier, but the charcoal used for tea is hard to come by locally. It is made in Japan from a variety of woods, each piece with a special dimension and requiring its own placement in the ash. In the West the convenience of a portable burner or single electric burner is more often than not a good substitute for heating hot water. For this occasion, I used a single portable burner.

> If the tea is held outdoors, charcoal, an alcohol burner, or even
> a camping stove can be used. If none of these are available, there
> is always the alternative of a suitable flask of hot water, preferably
> one with a spigot.

I found a spot for the table in the family room, near two large sliding doors that looked out onto the garden. The question arose of where to display a scroll and flowers. With no alcove in the room, what would be appropriate? An adjacent wall and side table proved suitable for hanging a print and displaying some flowers.

I had decided to use an old but sturdy wooden game table rescued from a friend's attic. The legs had been shortened so that the top was about two or three inches below my knees when I sat down on a chair. The surface was wide enough to make tea and accommodate a kettle and utensil tray. I brought out a small stand, lower than the table, and put that to the left of my place for the rinse water container. Next, I arranged three low stools for Jack, Patricia, and Anne. Facing the table, they would be on my right as is customary in the tea room. I also found wooden tray tables for their tea and sweets. On the day of the tea, I followed a simple plan of preparation:

> ☙ I began by setting a handmade ceramic brazier on a square piece of tile and centered the tile on the table. The tile was twelve inches away from the table's front edge, leaving enough room for the placement of the utensil tray. The electric burner fit neatly inside the brazier, and I was able to plug it into the nearby outlet.

> ☙ Forty-five minutes before my two guests arrived, I filled a small iron kettle with filtered water, set it down on the burner, and switched on the heat.

> ☙ Having chosen "autumn" as a theme, I hung on the wall a modern print that was reminiscent of an October dusk, with layered hues of lavender, crimson and saffron.

> ☙ On a wooden side table below the print, I placed a few sprigs of fresh purple asters, Japanese anemone, and eulila grass (sprinkled with water in the kitchen) inside the container of a light-brown basket woven in Indonesia.

> ☙ I went into the kitchen a few minutes later to assemble the utensils and sift the tea.

Sifting ensured the tea would dissolve easily and froth when it was whisked. Sealed in a small green can, the tea was a gift from Anne, who had purchased it in Seattle. She had generously included a round sifting tin found in the same store. This had a built-in, removable wire mesh screen near the top, somewhat like a conventional flour sifter. It also came with a small shiny spatula. The procedure for sifting is as follows:

🌀 Scoop some tea onto the screen and, using the blade of the spatula, sift the green powder across and through the mesh.

🌀 After all the tea has been sifted into the tin below, lift the screen off and carefully spoon some of the sifted tea into the caddy.

🌀 Following the traditional procedure, create a mound in the caddy that resembles a green mountain.

I decided to use two bowls, choosing the two I had bought from a potter in San Francisco. Washing and drying these for the occasion, I planned to bring one in on the round, black-lacquered tray, and the other with the rinse-water container. In the first, I would put in a dampened and folded tea cloth. The cloth was the standard *chakin*—a piece of white linen, eleven by six inches, hemmed on the longer edges. I rinsed it, wrung it dry, and folded it. Folding a tea cloth is simple:

🌀 Stretch the cloth lengthwise and fold it away from yourself in thirds.

🌀 Hold it vertically, then fold it in half over your left thumb.

🌀 Fold it in half once more.

🌀 Carefully remove your thumb and set the cloth aside on the rounded fold where your thumb was.

Making certain the front of the first bowl faced me, I put the folded tea cloth inside, flat side toward me, and rested the tines of the whisk against it.

Turning over the bamboo scoop, I placed it across the rim of the bowl, just to the right of the whisk.

When I first began to practice, I didn't own a tea scoop, so I used a long iced-tea spoon instead. The traditional bamboo tea scoop, or *chashaku*, is far more useful and tasteful an implement, and these days is not so difficult to obtain in certain Japanese stores. Carved out of a single piece of bamboo, it is generally about seven and one-half inches long, with a bent tip that is usually rounded. The scoop end is symbolically referred to as "dew," possibly because a few grains of powdered tea tend to cling to it when it is used. The makers of these scoops have their own way of carving the "dew" and handle, which make them unique and personal. A scoop is held by the end of the handle to avoid contamination of the tip. (The tip is touched with the silk napkin when purified. It is never touched with bare fingers.)

Because powdered tea is quite concentrated, two ample scoops (the equivalent of a full rounded teaspoon), with about two ounces of boiling water (less than a third of the bowl), are enough to make a bowl of tea.

After placing the bowl on the tray, I set the caddy at the 12 o'clock position on the tray, just above the bowl, and covered the tray with the *fukusa*.

Unlike the tea cloth, the *fukusa* is a doubled piece of silk that is specially sewn and is almost square—ten and a half by eleven inches. Traditionally, men use purple cloths, while women typically use red ones. Anyone unable to find a genuine *fukusa* can use a fine cloth napkin or a hemmed square of silk. In traditional tea, the cloth is folded in a precise way that requires training. It is folded the same way several times throughout a ceremony. This quiet action of refolding the smooth material is a skill that every host strives to learn in order to purify the mind as well as the utensil.

Taking the covered tray to the table, I centered it just below the steaming kettle, and returned to the kitchen to cleanse my ceramic rinse-water container and prepare the sweets.

Instead of traditional Japanese sweets made of flour and *an*, a sweet bean paste, I chose seven large, ripe green figs, which I grouped on a round gold-leaf plate. I added some white folded paper napkins, the customary *kaishi*.

☞ When presenting this tea, you may choose to serve a different sweet such as a moist cake or cookies of any kind that you feel would go with the taste of green tea. The sweets may be grouped on a large attractive platter or served on small individual plates for each guest. If no *kaishi* is available, other small folded paper napkins may be used.

Next, I prepared a basin of water and bamboo ladle for the guests. Not having a traditional *tsukubai* (stone basin), I found a deep earthenware bowl, took it outside my front door, and set it on a flat rock by a rhododendron bush. Then I filled it with cool water and left a hand towel in a small basket, next to the bowl on the rock. Taking scoops of water with the ladle, I dampened the rhododendron bush and surrounding plants, then trailed a wet path toward the front door to produce an inviting freshness—much as I would for a *roji*, the inner garden of a tea house. I drew more water and rinsed my hands and mouth as is customary, away from the bowl. This helped to refresh my mind.

The early September day was overcast. A pearl-grey light filtered through some thin clouds, so there was no need to turn on a lamp in the tea area. Instead, I lit a candle on a bronze stand in a far corner. My original idea was to have some flute music, but silence is sometimes preferable. Just as the doorbell rang, I tilted the lid of the kettle to allow some steam to escape and act as our music.

Figure 17

A table-style whisked tea

When Anne arrived, she noticed the earthenware bowl on the rock and ladled some water over her hands. After purifying her palms and mouth in the traditional way, she dried her hands and took a moment to enjoy the soft breeze and freshness of the rhododendron before ringing the doorbell. As I opened the door and Anne stepped inside, we saw Jack and Patricia coming up the walk. Jack guided Patricia in using the water basin. As the two of them joined us a few minutes later, Jack mentioned that having a water bowl ready had more than helped to empty his mind.

Both Jack and Anne knew from past visits where to leave their coats and knowing Patricia would be informed by Jack, I left them alone to relax and went into the kitchen. (If your guests are unfamiliar with tea, it is quite appropriate to guide them through the initial steps before the tea procedure begins.)

I had appointed Anne as first guest when I called to invite her, so she preceded Patricia and Jack to the side table where the painting on the wall and flowers were displayed. Each in turn bowed and stood for a while viewing the painting and flowers, then bowed again. Following the usual procedures of *chanoyu*, they took turns sitting in my place to look more closely at the bubbling kettle and the covered utensil tray. Once they sat in their places, I greeted them formally with a bow, and said that I looked forward to making tea for them.

I came in with the plate of fresh figs from the kitchen, which I placed before Anne on her tray-table, exchanging a bow.

Returning to the kitchen for the second tea bowl and rinse water container, I carried the bowl in my right palm and the rinse water container in my left hand, down by my side. I then sat down before placing the rinse water container on the short stand to my left, and the second bowl on the tea table far left of the utensil tray. Composing myself, I took a deep breath and focused on the rich texture of the red wiping cloth covering the tray.

I could sense Anne's keen attention. Well trained in tea, she had mentioned the importance of including the traditional folding of the wiping cloth in this adaptation. As far as she was concerned, folding the cloth and purifying the utensils went to the heart of tea ceremony. The mindfulness of tradition was maintained by using the customary method.

Carefully picking up the silk cloth from the tray, I held its top corners with my thumbs and index fingers and began the initial folding of the *fukusa* in order to purify the tea caddy and bamboo scoop.

To fold the silk cloth:

⟳ Fold the cloth in a triangle, holding the tips between your thumbs (on the front side facing you) and index and middle fingers (on the back).

⟳ Still holding the left corner between your thumb and index finger, shift the other three fingers of your left hand up to the side of the *fukusa* facing you; then bring your right hand up and over.

⟳ Drop the left corner, bring your left hand up and fold the hanging triangle in thirds lengthwise, holding the cloth between your thumb and left palm.

⟳ Turn the palm of your left hand up, allowing the cloth to rest on your left palm; use your right hand to fold the cloth down over your left thumb.

⟳ With your right index finger, draw an imaginary line from left to right across the top of the cloth and fold the ends of the cloth under your left hand.

⟳ Transferring the cloth to your right hand, fold it in half again with your left; remove your right index finger, then secure the cloth in your right hand.

The folding of a wiping cloth may seem unnecessarily slow or too deliberate, especially for a new student of tea. However, as Anne once mentioned, the purpose is to focus a guest's attention on the subtle dance of the hands and the colorful flow of the cloth, not on the host. With practice, the movements are managed with ease.

Figure 18

Folding wiping cloth:
parts 1 and 2

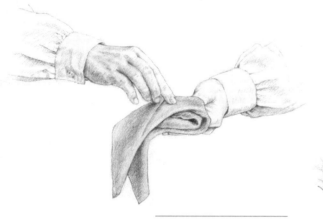

Figure 19

Folding wiping cloth:
parts 3 and 4

To purify the tea caddy:

☙ Hold the folded red cloth in your right hand and pick up the caddy with your left. Gently brush the cloth over the top half of the lid from left to right with a circular motion. Then wipe the bottom half of the lid from left to right with the same motion.

☙ Allowing one fold to open, wipe straight across the entire lid and close the fold in your hand. Holding the cloth in your right hand, set the caddy down with your left, this time on the upper left side of the tray at the eleven o'clock position.

☙ Place the folded cloth in your left palm, allowing the fold to open again.

Wiping the tea scoop deserves the same calm motions as the caddy, but first the cloth should be refolded.

☙ Turn the fold over, pick the top point and open it to its triangle, then fold the cloth as before.

☙ As the right index finger draws that imaginary line from left to right across the top of the cloth, fold the ends under your left hand.

☙ Without another fold, place the cloth on your left palm, pick up the slender end of the tea scoop, and put it on the center of the folded cloth.

To purify the tea scoop:

☙ Hold the cloth away from the body at waist level and above the knees. Keeping the scoop steady, fold the cloth over it and run the cloth straight out to the tip of the scoop with a light pinch-and-release movement. Then back, and out, wiping the scoop two more times.

☙ Place the scoop on the tray just to the right of the bowl at about the five o'clock position.

☙ Transfer the red cloth to your right hand and use it to close the tilted lid of the kettle.

☙ Set the still-folded cloth on the tray to the left of the bowl.

The choreography of movements and positions, from picking things up to setting them down, is extremely important. Unless movements are carefully considered, it is easy to become confused or hesitant when making a bowl of tea. How a utensil is handled is equally important, for this handling ultimately reflects the state of mind of the host and the respect given to whatever is in hand, whether it is a simple tea scoop or priceless antique caddy. As engrossing as procedures can be, however, the host must always be aware of the guest. A teacher once told me that when making tea, the host should make sure that "the guest never feels lonely."

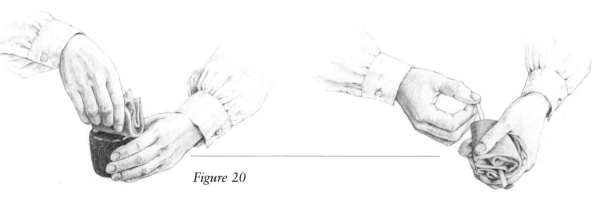

Figure 20

Wiping tea caddy and tea scoop: 2 part

Once the caddy and scoop are purified, it is time to prepare the whisk and the first tea bowl. Wetting and inspecting the tines of the whisk and rinsing and purifying the bowl are integral parts of the traditional way of making a bowl of tea. The gestures are not only calming but efficient.

To prepare the whisk and bowl:

☉ Remove the whisk from the tea bowl and place it next to the caddy at the one o'clock position. Then take the damp white tea cloth out and lay it on the tray to the right of the tea bowl at the three o'clock position.

☉ Pick up the red folded cloth from the tray with your right hand while you lift the kettle with your left and, securing its lid with the red cloth (fold turned over), pour hot water into the bowl—enough to fill less than a third of the bowl.

☉ Return the kettle to the burner, set the cloth back on the tray (fold side up) in its usual place (at nine o'clock) to the left of the bowl, then pick up the whisk and place it in the bowl, tines in water.

☉ Steadying the bowl with your left hand, lift the whisk out of the water, turning the whisk toward you to inspect the tines before setting it in the bowl. With right thumb on top of the handle of the whisk, lift the whisk out once more, turning it again as you set it back down in the bowl.

☉ Again, with the right thumb on top of its handle, pick up the whisk, swish the tines back and forth in the water of the bowl, making a pleasant sound.

⤷ Make a clockwise circle, draw out the whisk and allow the few extra drops of water to fall from the tines before setting the whisk on the tray at the one o'clock position.

⤷ Pick up the bowl with your right hand, transfer it to your left, and then empty the water into the rinse-water container.

Drying and purifying the tea bowl is relaxing and a pleasant part of traditional tea making.

To dry and purify the tea bowl:

⤷ Hold the bowl in your left hand, pick up the white tea cloth and place it inside the bowl, then pinch the folded top and lay it over the edge of the bowl.

⤷ Holding the bowl above your left knee, slide the cloth and bowl counterclockwise through your left hand three and a half times, letting the right hand meet the left at each turn.

⤷ With the last half turn, move the bowl in front of you, making sure its "front" faces you.

⤷ Lifting the cloth off the edge, put it inside the bowl and fold it over once.

⤷ Wipe the inside of the bowl—first the sides and then the bottom— and leave the folded cloth in the bowl.

⤷ Place the bowl down on the tray, and remove the cloth to its three o'clock position on the tray.

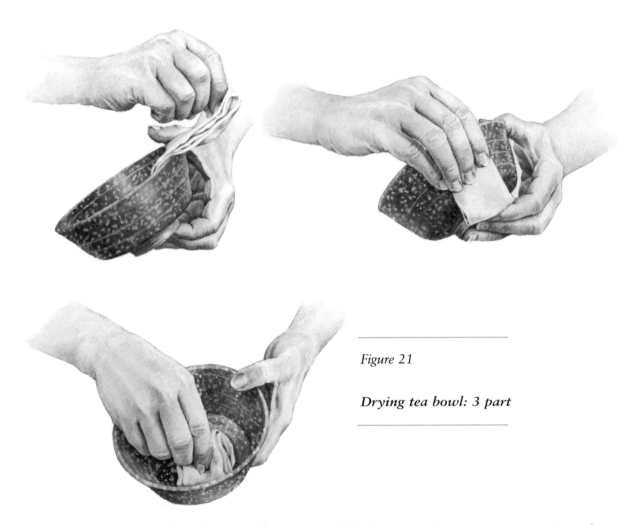

Figure 21

Drying tea bowl: 3 part

Now that the utensils were purified, it was time to scoop tea into the bowl. As I picked up the tea scoop, I invited Anne to help herself to the fruit. Following traditional etiquette, she politely turned to Patricia, excused herself for going first, and lifted the plate of fruit in thanks. After setting it down, she placed a luscious fig on a piece of *kaishi* paper before passing the plate to Patricia and Jack, who took turns helping themselves to the figs before returning the plate back to Anne.

While Anne enjoyed the fruit, I began making a bowl of tea for her.

To make a bowl of tea:

⊚ Pick up the caddy with your left hand and remove its lid with your right; put the lid down on the front edge of the tray, by the bowl.

⊚ With the tea scoop in your right hand, slightly tilt the caddy toward the bowl and spoon a full measure of the bright green tea into it. (Take care not to break the mound inside the caddy but scoop the tea from the side as if creating a valley.) Put in a second scoop of tea and lightly tap the edge of the scoop once on the right rim of the bowl to loosen any excess powder; replace the lid, set the caddy and then the tea scoop down on the tray in the usual positions.

⊚ Lift the kettle with your left hand and, securing its lid by using the folded red silk cloth in your right, add just enough steaming water to fill about a third of the bowl.

⊚ Replace the kettle and set the red cloth on the tray.

⊚ Steady the bowl with your left hand while you whisk the tea.

Figure 22

Whisking tea

133

Making Anne's tea, I used a gentle yet controlled forward and back whisking motion to bring the brilliant colored liquid to a smooth foam. Steam from the bowl carried the fragrance of the tea into the room. There was only the sound of the whisking, which seemed to bring an edge of calm that ordinary events seldom offered.

Before I set the bowl down for Anne, I turned it so that its front would face her. As it is customary for the guest to leave her or his seat and move forward to receive the tea, Anne walked over, took the bowl, and returned to her place. She moved the bowl to her left, toward Patricia, and bowed to her. Prompted by Jack, Patricia bowed with her and said, "please go ahead." Moving the bowl back in front of her, Anne thanked me, as we bowed together. She brought the three of us to the immediate present by noticing how the white-and-black-crackled glaze of the bowl was "a perfect contrast to the gorgeous green of the tea." She then raised the bowl and, turning the front away from herself, began to drink.

Once Anne had taken a sip, I made Patricia's tea in the second bowl. Meanwhile, Anne finished her tea and passed her bowl to Patricia so she could take a look at it. Before receiving her tea, Patricia set Anne's bowl down at my place. I then proceeded to make tea for Jack. When I rinsed Jack's bowl after he returned it, Anne offered me the dish of figs and asked if I would also have a bowl of tea. Since I had already sampled a fig that morning, I declined the fruit. In most traditional Japanese gatherings, the host prepares tea but does not drink with the guests; however, in this adaptation, I made myself a bowl.

Anne was a conscientious first guest and soon asked whether all of us would like more tea, but we were satisfied. When I rinsed my bowl, it was time to bring the tea to a close.

To close the ceremony:

☞ Place the second tea bowl on the side of the table left of the tray.

☞ Pour some hot water from the kettle into the other tea bowl on the tray, again steadying the lid of the kettle with the red folded cloth.

☞ Pick up the whisk with your right hand and steadying the bowl with your left, swish the water back and forth gently in order to cleanse the whisk. Lift the whisk out of the water, turning the tines toward you before setting it back down again. With the right thumb on top of the handle, make a clockwise circle before drawing the whisk out of the water and setting it down on the tray.

☞ Discarding the rinse water from the bowl with your left hand, pick up the white tea cloth with your right, place it into the bowl, and set the bowl back down on the tray with the right hand.

☞ Put the whisk inside the bowl with its tines resting against the tea cloth.

☞ Pick up the tea scoop with the right hand and red silk cloth on the tray and, holding the tea scoop in the last two fingers of the right hand, refold the cloth to purify the tea scoop. The tea scoop is held on the side of the cloth away from you.

☞ Purify the tea scoop, abbreviating the procedure by wiping the scoop only twice before turning it over and setting it across the rim of the bowl.

☞ With the folded cloth in your left hand, place the tea caddy in its original position at 12 o'clock on the tray.

☽ Dust the tea residue on the folded cloth into the rinse water container.

☽ Shift the folded red cloth to your right hand and use it to slightly tilt the lid of the kettle to allow the steam to escape.

☽ Unfold the red cloth and cover the tray.

☽ Take the rinse water container and the second tea bowl out of the room. (The bowl is held in the right palm and rinse water container in your left hand.)

☽ Come back into the room and remove the covered utensil tray, then return to remove the plate of sweets.

☽ Leave the kettle on the table.

Finally, reentering the room, I ended the ceremony with a standing bow. Anne, Patricia and Jack stood and bowed with me. Other than "adapting" the use of the red cloth to cover the tray, the entire procedure was much like *ryakubon*, the simplest presentation of traditional tea.

While I was in the kitchen, they took their turns and spent a minute or two looking at the painting and flowers before turning toward the tea table for a last glance at the cooling kettle. All three had plans that evening and left soon after. When they were gone, I cleared the tray-tables. Sitting down, I enjoyed the candle burning in the corner of the room and watched an evening shadow along the lawn outside the glass doors. I remembered the way Anne lingered remarking on the seasonality of the flowers, painting, and even the sweets. She was especially glad I had served figs, saying with a laugh that in recent months she had come to feel "too traditionally sugared by tea sweets." Patricia, who was unfamiliar with the taste of powdered tea, had been quite silent during the ceremony. Later, in a card I received from her she described how moved she was by the seeming simplicity of the ceremony and the "peace of it all."

As the afternoon began to fade and the wind blew large maple leaves across the grass, I heard the outside wind-bell ring, faintly at first, then clearly as it greeted the night. Cool as it was outside, I felt a definite warmth emanating from the iron kettle and from my guests' presence long after they were gone.

A Touch of Ceremony

As I devised different ways of serving tea, I became aware that whether at home alone or out in public, a tea gathering without ceremony limited the kind of removal I hoped to achieve. With those familiar with the basic principles, a feeling of calm was easily created. In theater, true removal from the outside world comes from a "willing suspension of disbelief." Tea, however, is not about performance. It is perhaps akin to a kind of suspension—physically and psychically—from everyday anxiety. Even in adaptations, I noticed that when my guests and I were totally present and focused, a kind of ease prevailed. Alone at home or in a workplace or restaurant, the key was to focus, practice, and be in the moment, even with a paper cup of ordinary bagged tea.

At the Workplace

Here removal is difficult, at times impossible to imagine. Yet it can be achieved whenever opportunities occur. There is no need to draw attention to ceremony. In fact, discretion is best especially for those colleagues who might not understand your purpose. Probably, all you can do in a short break is to bring deeper attention to the making and drinking of a cup of tea. In a private corner, free from faxes, phones, and interruptions, it is entirely possible to pause and raise a cup of tea a little higher than usual, in gratitude as we remember our interdependence in all things and respect for ritual.

A Group Tea

The idea of planning a group tea came when Anne and I met one sunny afternoon at a Japanese restaurant where hot towels and tea were offered. I arrived first. Knowing I could count on her to appear without any trite complaints about parking or politics, I went ahead and ordered our tea. A minute or so later she was sitting across from me, and, noticing the pot of green tea, she smiled a silent thanks. She arranged the cups neatly by the tea pot and asked whether she might pour. Together, we raised our tea before sipping the warm brew. It was an ordinary Japanese tea, but it gave a particular ease when shared in such a moment of quiet friendship.

After a comfortable silence, Anne asked whether I thought the spirit of tea came from the beverage or the sharing. It was a good question, but the waiter returned before I could say that the spirit of tea came from both the beverage and the sharing. After he left, I told her that trying to create a moment of removal with others in a public place raised problems. We then wondered about the possibility of a group tea, whether at a meeting or a small conference. How many people could be involved? When and where would such a ceremony be appropriate and practical?

Anne said there were few problems with two or three people, but with larger groups there might be "almost too many bodies and cups." At a business meeting, for example, the bigger the group, the more difficult it would be to draw the people away from the topic at hand for a break. However, the more we thought about it, the more convinced we were that a group tea could bring a sense of communality to a business conference, a workshop, or spiritual meeting. For instance, a tea ceremony could provide a moment of focused peacefulness before a business meeting involving negotiations, or after a discussion or a meditation at a retreat. Such a gathering could include tea followed by a few moments of quiet. The use of flowers or calligraphy would add ambience to the place but might not be feasible because of the difficulty of placing furniture and seating so many with limited space and time. Yet a feeling of calm and some aesthetic appreciation could be achieved even with only a minimum of planning.

Finishing the last piece of sushi, Anne tilted her tea cup and looked inside at the leaves. Smiling, she asked whether I could tell her fortune, though she knew I would not begin to try. I told her that once when I had visited Kyoto, I learned that the Japanese sometimes saved green tea leaves— especially those of *Gyokuro*, "the king of Uji tea." After brewing, these leaves were used in making a tempura batter or a type of chutney served with rice, or even used as a relish. Thus both the beverage and the leaves were consumed, as they are with powdered tea.

The waiter brought the check and two hot towels he had forgotten to bring when we first arrived. Once he was gone, we paid and each took a towel and wiped our hands and faces. Later at home, I looked at some of the notes I had taken of Anne's suggestions for a group tea.

The primary considerations of such an adaptation are privacy and seating arrangements. Ideally, the place would be large enough to accommodate a group seated in a circle or semicircle. Whether on the floor, in rows of chairs, or around a table, once the group is arranged and seated its members become guests. Depending on their number, a designated host will need one or two assistants.

The host plans and supervises the actual making of tea and the serving of any sweets, keeping in mind that green or herbal tea does not require cream or sugar and is easier to serve. For such occasions, cookies are best to offer, since cakes or pieces of fruit would require too much preparation and handling.

If guests are seated around a table, the surface has to be cleared. If they are seated in rows of chairs, the chairs can be rearranged in a circle or semicircle; or, if everyone is seated on a circle of cushions on the floor, the circle should be opened at one end for tea to be served.

At a table, guests would be served from the side. If they are seated in a circle of chairs or in a circle on the floor, the host must decide how to arrange the cups and sweets. A suitable place must be chosen for opening the circle and allowing the host and two assistants to leave, prepare the tea, and then come back and serve the group before joining them.

For an adaptation with a group of twelve, the host and two assistants may prepare:

    ⟡ Two large trays, each with six cups and six cookies on a napkin.

    ⟡ Two pots of tea, each with a folded napkin for pouring.

Before the tea begins, it may be wise for the host or leader to prepare the guests by explaining the etiquette and procedure of the tea. For a group seated on a circle of cushions:

    ⟡ Assistants enter the room one by one with the trays of empty cups and cookies and bow to the circle; those in the circle bow in unison. This signals the beginning of the ceremony.

    ⟡ The first assistant enters the opened end of the circle and, facing the leader of the group, bows and offers a cup and cookie, then continues to offer cups and cookies to the other guests.

    ⟡ The other assistant helps the first assistant in offering cups and cookies to the guests, remembering to place these items for the host and assistants as well, moving from the opposite end of the circle until everyone is served. The empty trays are placed behind the opened ends of the circle before the second assistant joins the circle.

⁙ The host enters with two pots of tea.

⁙ The first assistant takes one and helps the host pour the tea.

⁙ With everyone served, the empty teapots are set beside the trays.

⁙ When certain that everyone is ready, the leader brings two hands together and bows, and the guests do the same, palms together, heads lowered in respect. Then joining the leader, guests lift their cups in appreciation, before they begin to drink. There is silence during the taking of tea and sweets in this time of shared communion.

⁙ When everyone is finished, the leader again initiates a bow to signal that the tea has ended.

⁙ The assistants collect the empty cups while the host removes the teapots.

Such a ceremony might lead to a sympathetic and sensitive resumption of a meeting, or evoke a feeling of calm that would benefit even those in a business endeavor.

An Herbal Tea Adaptation

Some might question whether herbal tea has enough "substance" for ceremony, whether herbs contain "the essence of true tea." But herbal tea not only satisfies the tastes of many tea drinkers who seek variety, it also fulfills the needs of those who, for one reason or another, wish to avoid caffeine. In a sense, herbal tea is not unlike green tea as a source of nutrition and an instrument of healing.

The question of "essence" came up one day when I called to invite Jack for tea. Because of an illness, he was "off caffeine for a while." I suggested that we try an informal herbal ceremony, "perhaps a little blend of hibiscus and lemongrass." Jack agreed, amused at my solicitude. He liked the idea of herbs as a healing alternative to his antibiotics. "Herbs have a life of their own, and most are especially sensitive to infusion," he reminded me, "so much so that when making herbal tea you should avoid boiling water. Only hot water should be poured over the herbs, and very gently at that."

I followed these suggestions and, instead of choosing an Eastern scroll or a calligraphed piece of Western wisdom, blew up a large gold balloon that skipped across the floor and greeted him when he opened the door. He smiled, picked it up, and rolled it into a corner where we could see it. Planning to present a ceremony similar to what I did for English tea, I had set the table with two stacked tea bowls and a caddy next to the same white, candle-warmed teapot I had used before, filled with hot water. In the kitchen was another pot of hot water. Inside of it, the fragrant herbs were steeping. By each of our place was a small plate for the sweets I planned to serve.

When he sat down, Jack complimented the white columbine I had placed in a green ceramic vase on a smaller side table across the room. Before serving the tea, I excused myself, went into the kitchen, and returned with his favorite traditional Japanese sweets on a small black-lacquered tray. Called *nerikiri*, they were made with *an*—sweet bean paste—and resembled blue iris buds. He made a slight bow, excused himself for "going first," then raised the tray in thanks, removed a sweet, placed it on his plate and passed the tray to me. Raising the tray slightly in gratitude, I took one of the sweets before passing the tray back to him. The *nerikiri* were wonderfully delicious and came from a local Asian store in Portland.

After enjoying the sweet, I excused myself, picked up the white tea pot from its candle warmer, and left the room. In the kitchen I carefully strained the brewed herb tea into the tea pot and rejoined Jack, placing the pot back on its stand.

Pouring his tea, I placed the bowl in front of him. With his usual grace, he raised and turned it before taking a first sip. Finishing the tea, he was quick to compliment its warmth and taste and asked if he might pour me some. We spent some quiet minutes together, sipping tea, listening to a bird song outside the window as a fresh breeze entered, not wanting to break the peaceful spell with conversation.

After we had had sufficient tea, I took the tray of sweets from the table and left, then returned to remove the empty tea bowls. When I rejoined Jack at the table, we bowed together signifying the end of the simple ceremony. I left him alone for a while relaxing at the table as I packaged the remaining sweets as a gift for him.

Later, sitting comfortably on the couch in the living room, Jack emphasized that any tea was a satisfying communion, but he confessed that he was sorry that for him herbs would never replace the exceptional qualities of whisked green tea; he missed the sound of the kettle and the whisk. He then wondered whether a tea ritual could be considered a kind of "living theater," adding that by "theater" he meant that the guests were part of a performance. "Other than making tea and trading fine time, a host has no more need to create elaborate ceremony than he has to display dazzling utensils. If the gathering is harmonious, that's all that matters."

Later that evening, I thought about Jack's idea of "living theater." It confirmed my belief that the design of certain rituals did create a kind of theater. Nevertheless, the idea of "living theater" bothered me. If tea were "living theater," would it really matter where it was made, for whom, and, most important, for what purpose? This mattered to me, for I had learned long ago, in many places, people often seek too much theater and too little ritual.

Rituals create a feeling of intimacy through a gathering that unites people. The required focus of a ritual enables a momentary suspension of thinking—a time of "non-thinking," if you will. Participating in a ritual, something beyond the intellect is felt, something deeper in body and mind. The form itself as a way to intimacy depends on the individuals participating, on the focus each brings to it. However it is experienced, to some extent an inherent connection is felt by all involved. In this way the form of a ritual enlarges life and inevitably brings an experience of relief and in many instances deep joy.

A few days later, Jack and I happened to meet at a bookstore downtown. I told him about my concern with his "living theater." He admitted that without ritual, place, and some kind of ceremony, there would be no Way of Tea. "I guess if we've learned one thing, it's that tea is more than a beverage and the Way is more than a path. Of all the lessons in traditional tea, it seems that the important thing to remember is how tea can intensify a zest for living."

Tea Outdoors

The Way of Tea not only heightens awareness in everyday life but also produces a keener sensitivity to seasons, especially the turning of colors and the phases of the moon. Perhaps this is because certain traditional tea procedures change with the seasons. From May through October, the sunken hearth, or *ro*, in a tea room is covered and a portable brazier (*furo*) is used. This brazier is placed on a square tile or a plain or lacquered board and faced toward one corner of the room so that guests are not too close to the direct heat in warm weather.

As October nears, I am sometimes reminded of a haiku by Basho, in which he writes that autumn brings on an "inclination of mind" for a warm four-and-a-half-mat room. From November through April, the sunken hearth is uncovered and usually becomes the main source of heat in a tea room. If there is no hearth, a portable brazier can be used as a year-round alternative.

Even outside the tea room, tea gives us an intimate connection with nature. Somehow, those steaming cups of Safari Tea I once enjoyed in the Serengeti Reserve never tasted quite the same back in San Francisco. When a friend sent a pound of that Kenya brand to me years later, I made it last for several months. Though the flavor was good, it did not elicit the same sensations I had felt as I sipped from a tin camping cup and gazed at a stretch of brown-green plains.

We can have tea with whomever we want, with whatever formality we care to create. We can even pour a favorite tea into a thermos and take it on a day's hike. Wherever we are, tea can change the flavor and the pace of the day. It can promote a heart-to-heart communion between people at any time, any place, and sometimes under the most unusual circumstances.

I once heard a tale (which undoubtedly has many variations) about a monk who was captured by bandits. After robbing him of what little he had, the bandits told the monk, "You're going to die." Being a good Zen monk, he asked, "When?" The bandits replied, "Now!" Instead of panicking, the monk asked permission to make one last bowl of tea. The bandit chief thought this a unique request, and granted the monk's wish. Despite himself, the leader became fascinated with the monk's concentration as, in the face of death, he slowly prepared his utensils: his bowl, caddy, scoop, and whisk, purifying them with what little means he had. The monk boiled some water from his canteen and tapped powdered tea into his bowl. All these movements were executed with such focused attention to ritual and tradition that the monk soon mesmerized his executioners. They spared his life after he finished the tea.

This wise monk was probably carrying a conventional small box, or *chabako*, containing all the necessary utensils for making tea, except, of course, a kettle. A *chabako* can be taken anywhere to make tea indoors or out. It is designed to hold a bowl, caddy, scoop, whisk, and tea cloth, as well as a small container with tiny candies or pieces of colored, crystallized sugar as a sweet to enjoy with the tea. Most of these boxes are sold with utensils. Made in Japan, the boxes may be lacquered and in some cases are exquisite.

Every item in a traditional *chabako* is meticulously secured. In some, the tea bowl is encased in a silk brocade bag, and inside the bowl is a tea caddy in its own bag. The bamboo scoop is protected by another brocade pouch, while the whisk is stored in a wooden cylinder and the tea cloth in its own ceramic holder. The making of tea varies, depending on the utensils available and the occasion. When it is taken on a trip, the *chabako* may be securely wrapped in a cloth for carrying.

Finding boxes for tea utensils is not difficult. I found the one I use in a dusty retail warehouse in Portland. I have no idea what it was originally designed for, but the box, which I lined with black felt, had just enough space and height for my particular utensils. Into it I could fit a bowl with a container of tea wrapped in a silk cloth. I also put in a whisk, stored in its cylindrical canister, and a folded tea cloth neatly tucked into a carved bamboo holder Jack had made for me. The tea scoop was slipped into a narrow cloth pouch I had made. I eventually added another container suitable for sweet things to eat such as small pieces of candied fruit and nuts. Other practitioners make similar boxes to serve their particular purposes. Some utensil boxes are small enough to fit in a backpack or can be tied up and carried in square cloths.

One of my favorite memories is of a day hike I took with Anne, during which she made tea for us in a meadow beside a meandering stream. We had trekked for three hours and found our spot at the end of a steep trail that began at the base of Mt. Hood. After we rinsed our hands in the stream, I helped Anne heat our canteen water, using a tiny lightweight kettle and a Coleman stove we had brought along in our packs. The sky had cleared after a morning drizzle; sunshine filtered through the few remaining clouds. We stretched out a tarp to sit on, and Anne arranged her wooden box and utensils. She had a rugged bowl, a bamboo whisk in a plastic canister, a tea scoop, and a small container of powdered green tea.

While she made the tea, I sat comfortably, glancing from time to time at the tall firs surrounding the meadow. Using a clean bandanna as a wiping cloth to purify the scoop, she folded it and wiped the scoop with the same care and focus she showed in a tea room. The quiet rhythm of the stream and the movements of her hands dissolved all sense of time and sharpened every sensation. Perhaps it was the day, perhaps the green meadow, but I felt that that bowl of whisked liquid was prepared with such care that it unveiled the profound dimension of the mystique of tea.

Figure 23

Tea outdoors

8
Reflections

An adaptation is a valuable introduction to ceremony, but unless it is based on traditional principles, it is little more than a unique way of drinking tea. This realization brought me face-to-face with the differences between Eastern and Western sensibilities. I knew that each had advantages as well as disadvantages, but I learned that the tea ceremony offered positive aspects to both cultures. Guests entering a tea room need to consciously let go of outside anxieties. This letting go is a simple moment that more often than not defies description.

In the East, adherence to tradition is an obligation at most levels of society. Because life is rooted in tradition and ritual, tea ceremony is natural to the Easterner. Perhaps the difficulties Westerners have with the practice of ceremony stem from a rejection of conformity. In the West, where the cult of youth is predominant, tradition and, with it, the wisdom of elders, is often slighted or ignored.

To convey any tradition, the role of the teacher is crucial. The ceremony of tea is preserved by teaching and by the participants assisting each other. A teacher offers an opportunity in which one can polish the mind with practice, but the right to question a teacher is essential. Sound inquiry also demands faith and trust, for without trust, deep inspiration in any traditional practice will be difficult to achieve even with the wisest mentor.

There is a difference between inquiry and skepticism. With tradition, the secret is to guard against becoming too skeptical. If they have too many misgivings, students tend to use inappropriately what they are taught. If a teacher limits a student, however, it is up to that student to take in what is offered while preserving her or his own sense of inquiry. In this way, practice is enhanced. Basho's advice was not to follow "in the footsteps of men of old," but to seek "what they sought."

Along the Way, I have become aware of the living essence of tea through difficult and sometimes embarrassing effort. I have also discovered that tea provides not only solace, but also a new way to broaden perspective and clarify actions. In the intimacy of the tea room, the masks we wear and the barriers we sometimes construct fall away. The natural self is allowed an intimate connection with others.

My practice is sometimes complicated by conflicts between Eastern tradition and the Western promise of instant gratification. Life in the West is often fast-paced, whereas the reward of tea comes from slow, patient practice, from taking time to make a heartfelt bowl or cup of tea. Rikyu says that the object of practice is to ensure that a "delicious bowl of tea" is made. But a good bowl is not necessarily guaranteed by the quality of the tea. The Way is mastered as much with an open, quiet mind as with movements of the hands or the body. The spirit with which tea is made, received, and enjoyed is the key. This is because we do more than drink a bowl of tea—all of heaven and earth is present in the tea. Indeed, it is said that we drink the self and the cosmos through a bowl of tea. Such an experience can only occur if the entire mind and heart are open to the present.

However, tea is not a panacea or an escape. It offers no sudden enlightenment, only continual understanding. As Tantansai Sekiso, the fourteenth-generation Grand Master of Urasenke, said, "The Way is never exclusive. It is open to all to follow, but those who set out on the path . . . need the help of those who have passed that way before." [1]

[1] Soshitsu Sen XV, *Tea Life, Tea Mind,* page 80

Why tea? Why ceremony? Whenever I ask myself these questions, my mind wanders back to Singapore and the woman in the Chinese opera who offered a little girl tea in a tiny cup with no handle. Remembering that kindness makes me wonder whether the final question is not where I practice or how, or with whom. Beyond these is the question: Why *do* I make a bowl of tea? I believe the answer lies in purpose. By practicing tea ceremony as a means to open the mind and heart, something happens. The smallness of self-containment is lifted and a glimpse of the unbounded is revealed.

By mindfully making a bowl of tea, the self is free of everything but that clear presence in which tranquility exists. That anyone can share such presence with others is even more rewarding.

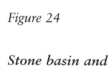

Figure 24

Stone basin and water ladle

Glossary

An

A sweetened bean paste generally used as a filling for soft cakes served with a bowl of powdered green tea.

Assam

A region in northeastern India, Assam is also the name given to teas produced in the area. Assam's famed tea plantations, including Thowra, Numalighur, and Napuk, yield black tea highly prized for its strong, full-bodied flavor.

Bamboo

A grass that grows mainly in warm or tropical regions. Some species grow as tall as one hundred feet. Bamboo stalks are usually round and jointed; they may be hollow or solid, with deciduous or evergreen leaves. Some types blossom only once in 60 or 120 years and then die. In the United States, the native bamboo is a cane. Among its many uses, bamboo provides wood, fiber, paper, and fuel. Long used for decorative purposes in gardens and art, in Japan it is also widely utilized in construction and in the making of tea utensils such as water ladles, tea scoops, baskets, vases, kettle-lid rests, boxes, and incense containers.

Basho, Matsuo (1644–1694)

A Japanese poet who brought the strict seventeen syllable form of haiku to an ingenious perfection. Turning to Zen Buddhism in his later life, he went on a spiritual pilgrimage. His poems reflect his travel and insights and evoke the mysteries of the universe as he strives to glimpse the eternal behind the transient life about him.

Bentinck, William Cavendish (1774–1839)

Governor-General of India; founded the Tea Committee in 1834 to promote the growing of tea in India. Among his many accomplishments in the interest of the people of India, he fostered communications and education, introduced reforms such as the admittance of Indians to higher office, and abolished *suttee*, the ancient rite in which widows commit suicide.

Black tea

A processed tea that has been withered, rolled, fermented, and dried before being sorted and graded.

Blended tea

A product of various teas blended to produce a certain flavor.

Bodhidharma

In Japan he is known as Daruma. The first Chan or Zen patriarch of China; he is believed to have left India for China in the sixth century to make a fresh start for Buddhism. The Chinese, more practical than speculative, took to Chan (or Zen as it is called in Japan), because it appealed to their down-to-earth logic. Eventually, the Chan school of Buddhism grew in China and spread to Japan where it took hold and is still a spiritual force in Japan's culture.

Brick tea

Tea molded under pressure into the form of bricks. These bricks are generally common grades of Chinese tea mixed with tea dust. Tribes in ancient China and Tibet are said to have used presses powered by oxen to crunch tea leaves into bricks to use for trade. In this regard, bricks came to be called "tea money." See Some Sources (Sur La Table).

Broken Orange Pekoe

A fine-quality black tea made from young leaves, including tips and broken segments. In India, this tea is considered of the highest grade.

Bruce, Major Robert

A Scottish adventurer who participated in the British conquest of Assam in 1824. Bruce lived among the native tribes of Assam and discovered that the tea they drank came from indigenous plants.

Camellia sinensis

Botanical name of the tea plant.

Celadon

Chinese porcelains with a translucent, pale green glaze.

Ceylon

The island now called Sri Lanka; its former name is still used in reference to the tea grown there.

Chabako

A tea box used for storing and carrying tea utensils. A *chabako* has all the required items for serving tea: a tea bowl, a tea caddy for powdered green tea, a tea scoop, whisk and a white tea cloth. Each item in the box has its own pouch or container for safe transport.

Chado

Japanese term representing the Way of Tea. It is a foundation of Japanese culture and can be traced to the ninth century, when tea was brought into the country from China. The study and discipline of *Chado* took hold in the fifteenth century, and its practice continues today. *Chado* encompasses not only the procedures for making tea, but also the history and philosophy of tea—the spirit of the Way.

Chai

A colloquial word for tea in India, as well as in the Middle East.

Chaji

A formal Japanese tea gathering, in which a fire to heat water for tea is laid and replenished, and a meal is served before thick and thin tea are offered. (See *koicha* and *usucha*.) The gathering lasts for about four hours and includes a short recess between the meal and the serving of tea.

Chajin

> Japanese term for dedicated practitioners of *Chado*, the Way of Tea.

Chakai

> A formal but abbreviated Japanese tea gathering at which only thin tea and sweets are served and sometimes a light meal. A *chakai* without a meal generally lasts about forty-five minutes to an hour.

Chakin

> A piece of white linen cloth, eleven by six inches, dampened, folded, and used for drying the tea bowl. For the serving of thick tea, or *koicha*, guests generally bring with them a smaller, dampened *chakin* for wiping the lip of the tea bowl after drinking.

Chanoyu

> Literally "hot water for tea," *chanoyu*, is sometimes used interchangeably with *Chado*, or the Way of Tea, the spiritual path that is the foundation of the practice of *chanoyu*. *Chanoyu* has social, aesthetic, and ethical aspects that bring people together to create peace and intensify the experience of life.

Chasen

> Japanese bamboo whisk. It is used for whipping or kneading powdered tea after hot water is added to a tea bowl. Because of their delicate tines, whisks are usually dampened in warm water before they are used. The standard height of a *chasen* is five inches; the tines are three inches long and the handle is two inches. Fashioned from a single piece of bamboo, the whisk's handle usually includes a node and is slightly less than an inch in diameter. The front of the whisk is where the black cotton thread separating the tines is knotted, and this is always turned upward when the whisk is placed inside a tea bowl. Although a *chasen* may differ according to the type of bamboo used or the number of tines, its basic design has remained unchanged throughout the centuries.

Chashaku

A thin, delicate-looking scoop used to measure powdered tea from the tea caddy into the bowl. The *chashaku* is carved out of a length of bamboo. In early times, ivory or metal was used. (Ivory scoops are still employed when Chinese utensils are used to prepare and serve tea.) These days, in some innovative circles, blown glass, steel, and even plastics are being used to make these scoops. The more prized tea scoops, however, are those carved by early tea masters. These early masters gave poetic names to their scoops, such as "Homecoming" by Senso, the fourth-generation Grand Tea Master, and "Spreading Fortune" by Joso, the fifth-generation Grand Tea Master.

Congou

A term used for all black teas from China.

Darjeeling

A northern province in India, Darjeeling produces some of the world's finest black tea.

Dust

The siftings of the tiniest particles of black tea.

Fannings

The siftings of better-grade teas, somewhat grainy particles slightly larger than dust.

Fermented tea

Black tea, in which the natural enzymes in the leaves are allowed to oxidize before drying.

Flush

The shoot of the young tea leaf. The term sometimes refers to seasonal harvests—for instance, the "first flush" in early spring, the "second flush" in late spring or early summer.

Fortune, Robert (1813–1880)

An adventurous British botanist and one of the first Westerners to discover that green and black tea come from the same plant and how it is processed produces the different colors. Bringing China's carefully guarded secrets about tea production to the West, Fortune enabled the East India Company to cultivate and process tea in India successfully.

Fukusa

Japanese term denoting a square piece of cloth, about ten and one-half by ten inches, usually of a fine, slightly heavy silk, used for purifying the tea caddy and scoop (and, for women, for holding the hot lid of a kettle) during a Japanese tea ceremony.

Fukusabasami

A flat, fabric purse used to carry a *sensu*, *fukusa*, *kaishi*, and other related effects for a Japanese tea ceremony.

Furo

A brazier upon which a kettle sits for heating water. *Furo* also denotes the summer season of tea (generally May through October 31st). During this season the brazier in the tea room is placed at a far corner away from the guests so that the heat is not directly felt by them.

Furyu

In Japanese, *Fu* means "wind" and *ryu* means "to flow," as in allowing the spirit to flow through life as the wind flows naturally through all of nature.

Gengensai Seichu (1810–1877)

At age ten, Gengensai Seichu was adopted from the family of a lord by the Sen family and grew up to become the eleventh-generation Grand Master of Urasenke. His education included the study of the literary classics of Japan and China, *Noh*, calligraphy, flower arrangement, incense, and poetry. Bringing all these gifts to his role as grand master made Urasenke a center of culture in Kyoto.

Golden Tea Room

The original Golden Tea Room of the warlord Toyotomi Hideyoshi was especially designed for serving tea to the emperor. Hideyoshi conceived of this idea after a visit to the emperor's court in 1585. To honor the emperor and to emphasize his own exalted position, Hideyoshi wished to impress the court with his portable Golden Tea Room. The gilded three-mat room had crimson tatami and gold-embroidered borders. There was a stand of gold-lacquered wood for the utensils, and all the utensils except the whisk and *chakin* (the white tea cloth) were of solid gold. In Hideyoshi's time, gold was both plentiful and popular in Japan. Many castles used gold leaf for decoration, but only Hideyoshi was bold enough to gild a tea room—even at a time when Rikyu, his acclaimed tea master, declared that tea should be kept simple and humble. The tea room was destroyed when Hideyoshi lost his power, but records of its construction were kept. When these records were uncovered in 1979, a replica was built and installed in the MOA (M. Okada International Association) Museum in Atami, Japan.

gong fu tea

A Chinese form of tea ceremony using the diminutive Yixing teapots and cups, believed to have been developed during the middle of the Ming dynasty and favored by connoisseurs of tea.

Green tea

Unfermented tea. After they are picked, the tea leaves are immediately heated to kill enzymes and prevent fermentation. The leaves are then rolled and dried.

Gunpowder Pearl Pinhead Green

This tea is hand-rolled into tight little pellets resembling old gunpowder. It is a fine tea that, when brewed, turns a light amber color with a touch of green, offering a delightful fragrance and a slightly sweet taste. Grace Rare Tea, of Grace Tea Company, Ltd., New York, touts it as a tea for contemplation and reflection.

Gyokuro

The Japanese term *gyoku* is generally translated as a precious stone in the shape of a ball, and *ro* in this instance means "dew." Thus the tea is sometimes referred to as "Precious Dew." The leaves are easily recognized, as they resemble pine needles, flat and sharply pointed. It is considered the finest of Japan's teas and is among the most expensive.

Haiken

The time during a traditional Japanese tea gathering when guests request to have a closer look at some of the utensils used in the making of tea such as a tea caddy or tea scoop. The word *haiken* also means the actual activity of examining the utensils.

Hideyoshi

See Toyotomi Hideyoshi.

Higashi

Japanese "dry" tea sweets, usually made of press-molded rice flour and sugar, served with thin tea. See *Namagashi and Okashi.*

Hishaku

A water ladle made out of dried bamboo, the handle of which is over thirteen inches long. In Japanese tea ceremony, the *hishaku* is used to ladle hot and cold water. A different ladle is used outdoors with the *tsukubai,* or water basin. It has a slightly larger scoop and is placed across the basin.

Ichigo ichie

A Japanese phrase meaning "One time, one meeting," often used in tea gatherings. It is a reminder that each moment should be treasured, because the same moment will never be repeated.

Jaku

Japanese word for tranquility. It is one of the four principles in the Way of Tea established by Rikyu.

Jin-chu

A Chinese green tea produced from tea plants exposed to more sun than usual. The long, slim leaves of these plants are light in color—almost white—and make an excellent brew, flavorful and robust.

Kaiseki

A light Japanese meal, sometimes served in courses to guests in a traditional tea gathering before tea is prepared.

Kaishi

Thin, rectangular pieces of white, handmade mulberry paper, approximately seven by five and three-quarter inches, folded in half. These pieces of paper, which come in packets of approximately thirty sheets, are generally used for holding Japanese sweets—much like a plate or napkin.

Keemun

A Chinese tea. Considered a wonderful evening tea, this fine black tea comes from the Anhwei province of China.

Kei

Japanese for "respect." The second of four principles in the Way of Tea.

Kensui

A Japanese container for waste-water, often made of copper, ceramic, or bronze.

Kimono

The Japanese kimono is made from cloth of various fibers, weaves, motifs, and colors. Wrapped about the body, its beauty is in the cloth and the way it drapes. An essential part of the Japanese wardrobe, even in modern life, it is a classic attire for tea ceremony, since it lends a definitive grace to the experience for both women and men. Accentuated by its stiff *obi*, or long waist-band, the *kimono* produces an elegance of movement and posture.

Koicha

> The thick tea served in a traditional Japanese ceremony. The procedure for making *koicha* differs from that for thin tea, or *usucha*, and is considered more formal. The tea used is of a finer grade than that selected for thin tea. Before *koicha* is made, guests have already been served sweets, such as those made from *an*, or red-bean paste. Three scoops of powdered tea per guest are put into a bowl. Hot water is added, and the whisk is used to knead the tea into a thick and smooth consistency. One bowl may serve up to five guests. Each guest takes approximately three and a half sips and then cleans the lip of the bowl before passing it on.

Kombucha

> In Japanese it means "seaweed tea," but in the West *kombucha* has come to mean a special tea made by fermenting a white, gelatinous culture of yeast and bacteria in sugared black tea. This culture is likened to a "mushroom" that reproduces itself as it ferments. *Kombucha* is believed to be a cure-all. A folk remedy in Asia for over two thousand years, the culture is rich in enzymes and Vitamin B.

Koshikake machiai

> The outdoor arbor in the tea garden where guests wait before entering the tea room. A bench with round straw mats is provided. Sometimes a tobacco tray is offered. This will have a pipe, tobacco, a slim bamboo holder with a little water, and a ceramic bowl that contains a beautiful mound of fine ash with a lit piece of charcoal. These days, the tobacco tray is more of a traditional offering rather than used for smoking. On cold days, when passing the tray from one guest to another, placing the hand over the lit ash bowl is a welcomed warmth. Also see *Machiai*.

Lapsang Souchong

> A Chinese tea favored for its smoky flavor, which is produced by drying the tea leaves over a wood fire—perhaps cypress or pine logs.

Lu Yu (725–804)

Born in Hupei Province in Ching Ling, now called T'ien Men, Lu Yu came to be known in Chinese history as the "god of tea." Famed for his *Cha Ching*, or *The Classic of Tea*, Lu Yu was the first to teach how to appreciate tea and how to lay out the necessary utensils for brewing.

Machiai

The Japanese term for the indoor waiting area for guests arriving at a tea ceremony. The *machiai* sometimes includes a tatami mat room and an area where sometimes a scroll may be displayed, along with a tobacco tray. See *Koshikake machiai.*

Matcha

Powdered green tea, produced in Japan from high-quality tea leaves and used mainly in tea ceremony.

Mizusashi

Japanese term for "cold water jar," one of the utensils used in a traditional tea gathering. The *mizusashi* can be of ceramic, wood, or metal.

Mizuya

The preparation room (a tea kitchen) usually adjoining a traditional Japanese tea house or tea room. This is where utensils for tea are readied.

Murata Shuko (1423–1502)

A Zen priest, born in Nara. He entered the Shomyo-ji temple at the age of eleven and eventually attained priesthood there. Shuko is considered the father of the tea ceremony, credited with originating much of the spirit and etiquette of tea.

Mushanokojisenke

Another school of *chanoyu* in Kyoto, founded by Sen Soshu (1593–1675), who named it after the street it was on—Mushanokoji. Soshu was one of three sons of Sen Rikyu's grandson Sotan.

Namagashi

Moist sweets, usually made of rice flour with a sweet bean-paste filling. The small cakes vary in shape and color, often resembling seasonal motifs. See *Higashi and Okashi.*

Natsume

A caddy for powdered tea, resembling the shape of a jujube fruit. The *natsume* is usually a finely lacquered container, although there are also beautifully crafted natsume of various woods that are not lacquered.

Nijiriguchi

The raised entrance for guests to a traditional grass hut style tea room. It is a two-and-a-half-foot-square sliding door that encourages humility by requiring the guest to enter the room on hands and knees. The elderly or those unable to kneel may use another entrance. (The host's entrance, the *katoguchi*, requires the head to be lowered when carrying utensils into the tea room.)

Noh

A dramatic art created in the fourteenth century where a tale unfolds in the slow, skilled movements of masked actors accompanied by exceptional chanting.

Okashi

Tea sweets served in a traditional gathering. See *Higashi* and *Namagashi.*

Omotesenke

A school of *chanoyu* founded by Sen Sosa, Sen Rikyu's great-grandson. Sosa named it after the front part, or *omote*, of his father's property.

Oolong tea

Oolong, meaning "Black Dragon," is a semifermented tea from China and Taiwan made from leaves that are allowed to oxidize partially before drying. The degree of oxidization varies, producing light, medium, and heavy varieties.

Opium Wars

The conflicts between Great Britain and China that occurred between 1839 and 1842 when China, attempting to stop the import of opium, seized and destroyed opium belonging to British merchants in Canton. Seeking to end Chinese restrictions on foreign trade, especially on opium, the British used the incident as an excuse to attack China's coastal cities. With more powerful modern weapons, they eventually defeated China.

Orange Pekoe

Black tea produced from certain leaves of the tea plant measuring eight to fifteen millimeters long. The flowery Orange Pekoe leaves are harvested early as unopened buds. The word *pekoe* (in Chinese, *pak-ho*), which denotes the fine hair on a newborn, refers to the light down found on young leaf buds. It is believed that the term "orange" may have originated with Dutch merchants who, wanting to convey the nobility of the tea, named it after the Princes of Orange. The earliest harvested tea yields a fine beverage, clear and very fragrant. Orange Pekoe is a slightly later harvest. Its longer, elegant leaves also make an excellent brew—light, delicate, and as flavorful as the younger pekoe.

Otsume

One who is designated as the "last" guest in a Japanese tea ceremony. The *otsume* is as experienced as the first guest (if not more) and assists the lead guest and host as needed.

Otherness

(See Removal)

Pai-yun

"White Cloud," a green tea from China noted for its distinct flavor.

Pandanus jellies

Thin green pieces of jelly that are added to coconut cream and palm sugar in a sweet shaved-iced drink called *chendol* served mainly in Malaysia and Singapore. Pandanus jellies come from the plant of the genus Pandanus. Called *Pandan* in Malay, it has sword-shaped leaves arranged in a spiral, as in screw pines of the same genus.

Pekoe

The label given to top grades of black tea. The word *pekoe*, meaning "white hair" in Chinese, originally referred to young leaves covered with down.

Powdered tea

Tea produced by pulverizing unfermented green tea leaves to create a fine, light powder. See *Matcha*.

Removal

A way to become detached from the trials and limitations of ordinary life that inhibit the awareness of the present. An experience of removal can be created through a ceremony in which there is special attention to detail and focus. The subsequent "letting go" of extraneous thoughts can produce a feeling of peace and contentment. This kind of removal also allows the self to contact *otherness*, that climate between concern and equanimity.

Robiraki

A tea ceremony celebrating the opening of the *ro* or hearth season.

Ro

The sunken hearth in a tea room. *Ro* also designates the winter season of tea, generally through the beginning of November to the end of April. It is a time when the hearth is opened and charcoal is laid for the kettle to heat hot water for tea. The hearth's placement in the tea room enables the guests to feel the warmth of the heat. The tea gathering called Robiraki celebrates the opening of the *ro* in early November.

Roji

> The landscaped area surrounding a tea house. The *roji*, or "dewy path," is divided into an outer and inner section and is designed to resemble a mountain path. Indigenous plants are selected—perhaps maple, pine, azaleas, ferns, and moss. Stepping stones are carefully chosen and laid for easy walking, leading the guest from the garden gate to the tea house. A stone lantern usually stands by the *tsukubai*—a stone basin filled with water and used by guests for purification.

Ryakubon

> The simplest form of Japanese tea ceremony. It uses a tray, tea bowl, *chakin* (white tea cloth), *chasen* (whisk), *chashaku* (tea scoop), *natsume* (tea caddy) filled with *matcha* (powdered green tea), and a pot of hot water.

Ryurei

> A procedure whereby the host, seated on a stool, makes tea at a table for guests, who are not seated on a tatami floor but on benches, stools or chairs. This procedure was created in 1872 by the eleventh-generation Urasenke tea master, Gengensai, when he served tea to Western guests.

Sabi

> The state of naturalness from which spontaneous creativity arises. Unlike *wabi*, which is a mood that can be created, *sabi* is natural activity, the kind that springs from a genuine acceptance of things as they are. Likened to the spirit of Zen, *sabi* can be seen in all parts of life, expressing itself in the wonder and beauty of various arts as well as in everyday activities.

Sabie Zen

> Originated by a group of professionals and artists of the Sabie Cultural Institute in Kyoto, this fresh perspective on the Way of Tea still maintains the original precepts of *Chado*, thus preserving tradition while encouraging innovation. In both East and West, Sabie Zen has inspired dramatically redesigned tea rooms, as well as modern tea tables, ceramics, utensils, and painting.

Samurai
A member of the hereditary warrior class in feudal Japan.

Sei

Purity, the third principle of the Way of Tea.

Semifermented tea
See Oolong tea.

Sen Rikyu (1522–1591)
Japan's most highly regarded tea master. (In Japan, the surname, in this case Sen, comes first.) Rikyu was not his given name, but an honorary one. Rikyu was best known for his innovative application of the concept of *wabi* to *chanoyu*. Through the principles of harmony, respect, purity, and tranquility, he influenced the direction of the Way of Tea. As a Zen practitioner, Rikyu sought unity with humanity and nature through tea. He felt that a mind free and present and a completely open heart were the prerequisites for entering a tea room, for only then could a real exchange and a meeting of minds occur. To Rikyu, the smaller the tea room, the better it was for creating such intimacy. He brought a freedom to the Way of Tea, a manner that enabled him to deal with any unforeseen problems.

Sen Soshitsu XV
The fifteenth-generation tea master of Urasenke, Dr. Sen is a descendant of Sen Rikyu. He is now called Sen Genshitsu, having relinquished his name of Soshitsu to his son Zabosai, who has become the sixteenth-generation Grandmaster of Urasenke. Dr. Sen is distinguished by many honors from his country and around the world for his promotion of "peace through a bowl of tea."

Sencha
A green tea made of the whole, unrolled (or natural) leaf of the tea plant. *Sencha* also refers to a brewed-tea ceremony practiced in Japan that has its roots in the traditional Japanese ceremony dating back to

the seventeenth century. The etiquette of *sencha* somewhat resembles the Chinese *gong fu* ceremony in its use of small ceramic pots and cups. Fine green tea is steeped in a small pot of hot water. The water is initially heated over a ceramic brazier and then cooled in a ceramic "boat." The brewed tea is poured into warm, dried cups placed on small saucers. These saucers can be of metal, bamboo, or other wood. Three different brews from the same tea leaves are enjoyed—the first generally referred to as "sweet," the second "bitter," and the third "astringent."

Sensei

This Japanese word means "teacher."

Sensu

A folded fan used at a Japanese tea ceremony. As a sign of respect, a guest places the fan in front of him or her when greeting the host or another guest, or between the guest and whatever object she or he is viewing upon entering and just before exiting the tea room. The fan is also used as a kind of place card that marks the guest's place in the tea room.

Shen Nong

Known as the Divine Cultivator or Healer, Shen Nong was the second of three mythical sovereigns. It is believed that he lived in 2737 B.C. and was born having the head of a bull and the body of a man. By inventing the cart and plow, yoking the ox, taming the horse, and teaching his people how to clear the land with fire, he established agriculture in the land. Many remarkable tales told of his youth mention that he began to speak three days after his birth and was able to walk within a week, and to plow a field at only three years of age.

Shogun

A military ruler in Japan.

Shokyaku

The main, or lead, guest during a traditional Japanese tea ceremony. The term suggests one who is experienced and can be emulated by the other guests. The *shokyaku* leads the other guests in entering and exiting the tea room and acts as their spokesperson.

Shou-mei

Chinese for "Old Man's Eyebrows," a green tea from China that is minimally processed to retain the scent and color of fresh-picked leaves.

Steps of Yu

Also known as the Pace of Yu. A shamanic dance that transported the legendary Yu to the stars. An extraordinary chieftain, Yu was known for having power over the elements. Performed by generations of Taoist priests, mystics and shamans, the dance of the Steps of Yu is also a practice of the internal martial arts.

Sutras

Similar to scriptures as in those pertaining to the sermons of the Buddha Shakyamuni.

Takeno Jo-o (1501–1555)

Influenced greatly by Murata Shuko's Way of Tea, Takeno Jo-o distinguished himself as a great tea master, enriching the forms and accessories of the art of tea. A poet as well as a Zen Buddhist, he later became a priest when years of war brought him face to face with the impermanence of life. His simple, quiet elegance in the practice of tea—a *wabi* concept—made a great impression on his student Sen Rikyu, who furthered that ideal in his own practice of tea.

Tana

A stand placed in the tea room for utensils used for making tea. The stand, made of wood, comes in different designs according to the season and the formality of the gathering.

Tanaka, Sen'o (1928–)

Grandchild of the founder of the Greater Japan Tea Ceremony Society. He became its president in 1961 after receiving his master's degree from Waseda University. The author of *The Tea Ceremony*, Sen'o Tanaka also publishes a monthly magazine, *Chado no Kenkyu* (Studies in the Tea Ceremony).

Tantansai Sekiso (1893–1964)

The fourteenth-generation Grand Master of the Urasenke School of Tea. He greatly increased the popularity of *chanoyu*, underwent Zen training from the head priest of Daitokuji in Kyoto, and became the head of Urasenke when his father died. During his career, he saw many changes in the political and cultural climate of Japan. His sensitivity to change led him to the idea of introducing *chanoyu* to the West. Restructuring Urasenke, he developed the International Chado Cultural Foundation, with study centers in Kyoto and Tokyo, and in 1962 established a full-time school of *chanoyu*.

Tatami

From the Japanese verb *tatamu* which means to fold or layer. Today, *tatami* generally refers to the floor furnishings of a Japanese home or tea room. The tatami consists of a base, a covering, and cloth borders. The base is of rice straw that has been dried over twelve months, stacked in several horizontal and vertical layers, and tightly bound. The covering is made of woven candle rush or *igusa* stalks, which measure roughly two and a half millimeters in diameter. The stalk's hard outer surface encloses a supple, spongelike core that cools the plant and at the same time absorbs and releases its moisture. These qualities of the plant are what make tatami so wonderfully cool to the touch in summer and warm in winter. The cloth borders around the tatami differ according to the room. Dark-blue or black cotton or linen borders are used in the tea room proper, while tea room alcoves may have colored, patterned borders. The size and quality of the tatami, and even its base, vary according to its use. In a tea room, the usual size of a tatami is three feet wide and six feet long.

Tea cake

Green leaves that are simply dried, crushed, and packed in the form of a cake. In ancient times, a small piece was cut off before infusion in hot water, and the leaves were shredded and ground into a powder with a special mortar and sifter. Sometimes the leaves were shredded into powder between fine pieces of paper. Earlier mortars of the Tang dynasty were usually made of wood, although a later Tang dynasty mortar of silver with gilding was found among recent discoveries at the Famen Temple in Shaanxi, China.

Tea gardens

Parks in eighteenth-century London that provided mainly amusement and areas where tea was served. (Not to be confused with the Japanese garden surrounding a tea house, or the gardens, estates, and plantations in East India where most of the world's tea is grown.)

Teishu

Japanese term denoting the host in a traditional tea gathering.

Temmoku

Japanese name for a type of Chinese ceramic bowl imported during the Song dynasty. It has a narrow base and a broad rim and is famous for its glaze. One in particular, the *Yohen Temmoku*, that resembles a star design is prized for its opalescent blue-green glaze with scattered silvery spots. It is considered a national treasure in Japan.

Tencha

Green tea leaves that are pulverized to make the powdered tea *matcha* used in the Japanese tea ceremony.

Tocha

Tea-tasting contests once held in Japan. Competitors exercised their expertise in distinguishing and judging various blends of tea.

Tokonoma

An alcove in a tea room or house, usually decorated with a scroll of calligraphy or painting. Sometimes a vase of flowers or an incense container is also displayed. Alcoves vary according to the construction of the house or the room.

Toyotomi Hideyoshi (1536–1598)

Born in humble circumstances, Hideyoshi was a military genius who eventually succeeded the warlord Oda Nobunaga as the military ruler of Japan. Hideyoshi was attracted to *chanoyu*, especially the *wabi* influence of his tea master Sen Rikyu. Hideyoshi profited greatly from the calming effects of his practice, but he also tended to use the tea room as a subtle means for conducting strategy meetings with his vassals and sometimes discussing confidential military matters with his officers or generals.

Tsukubai

A stone basin filled with cold water, used by guests to purify themselves before a tea ceremony. The term *tsukubai* can include the area around the basin.

Urasenke

The school of *chanoyu* started by Sen Soshitsu, Sen Rikyu's great-grandson.

Usucha

A "thin" tea offered during a traditional Japanese tea gathering that is less strong than thick tea, or *koicha*, and has a different texture. To make thin tea, two full scoops of powdered tea are placed into a bowl, hot water is added enough for three good sips, and the tea is lightly whisked into a froth.

Wa

Japanese word for harmony, the first principle of the Way of Tea.

Wabi

A Japanese word that can mean many things, such as simplicity, imperfection, solitude. *Wabi* was also defined by D. T. Suzuki as "transcendental aloofness in the midst of multiplicities."

White tea

A light, semifermented tea.

Zen

Zen, in its simplicity, means living in the moment-to-moment stream of life with complete awareness. Living Zen requires setting the past and future aside and penetrating each arising moment in the midst of everyday activities. Alan Watts felt that Zen was "a means of liberation" that dealt not with what was good or bad, "but what is." In her book *Zen, Direct Pointing to Reality*, Anne Bancroft said that Zen is not concerned "with the idea of Buddha or God, but with the reality of the human being," because "the true human being does not strive for what he can get out of life, but for what life is in itself, and he lives according to this knowledge."

Some Sources

For Tea and Utensils

Anzen Importers
736 Martin Luther King Boulevard
Portland, OR 97232
(503) 233-5111
Anzen Importers sells foods, kitchenware, and other goods from Japan, including a selection of green tea. Some traditional tea sweets from Sacramento and Los Angeles, California, may also be purchased.

Asakichi Antiques & Arts
1730 Geary Boulevard
Japan Center
San Francisco, CA 94115
(415) 921-2147
Offers some fine traditional Japanese tea utensils, powdered green tea, and standard antiques and art objects.

Dean & Deluca
560 Broadway
New York, NY 10023
(800) 221-7714
In addition to offering teas, this place is also a source for Japanese tea whisks and teapots.

Den's Tea, Inc.
2291 W. 205th St., Unit 101
Torrance, CA 90501
Toll Free: 1-877-DENSTEA (1-877-336-7832)
Imported directly from their factory in Shijuoka, Japan, Den's Tea offers not only powdered tea, but good qualities of brewed tea as well,

including Sencha. Also available are various utensils such as tea pots, bowls, and cups. Ask for their brochure.

Felissimo Tea Room
10 W 56th Street
New York, NY 10019
(212) 247-5656

On the top floor of a limestone townhouse-turned-shop, this tea room and art gallery offers tea in tiny stoneware pots, with handleless cups, for those wanting a rest from New York's busy streets. Delicious sandwiches and desserts are available with afternoon tea. One of the specialties is Osenbei cookies. The tea room may be reserved for special gatherings and receptions.

The Felissimo also offers a tea kit for those interested in *Chado*, the Way of Tea. This kit comes with Okakura Kakuzo's *The Book of Tea* and includes a ceramic tea bowl, bamboo whisk, tea scoop, linen cloth, powdered green tea, and sugar candies, in a kiri wood box.

Grace Tea Company, Ltd.
50 W 17th Street
New York, NY 10011
(212) 255-2935

Sells excellent rare teas, among them Superb Darjeeling 6000, Winey Keemun English Breakfast, Formosa Oolong Supreme, Before the Rain Jasmine, and Gunpowder Pearl Pinhead Green. The Grace Tea Company, a family business, states that it "hand-blends and hand-packs" its teas.

Imperial Tea Court
1411 Powell Street
San Francisco, CA 94133
(415) 788-6080

A fine tea room with rosewood paneling, dark furniture, and patterned green fabric on the walls. An excellent selection of Chinese teas is available and tea tasting is encouraged. (Customers may select a simple tasting or a more formal one.) On one of Chinatown's busy streets, the Imperial Tea Court offers a quiet respite where one may linger over a pot of specially brewed tea. The owners' display of unusual antique and recent teapots is an added attraction.

Ito En New York
822 Madison Avenue
New York, NY 10021
Ito En Tea Store: (212) 988 7111
Kai Restaurant: (212) 988 7277
www.itoen.com

This delightful tea store and restaurant is tucked away in the midst of busy Manhattan. It offers a fine selection of teas, teaware, wonderful *kaiseki* food, as well as an opportunity to taste *matcha*, the powdered green tea.

Jade Garden Arts & Crafts Company
76 Mulberry Street
New York, NY 10013
(212) 587-5685

Among its many arts and crafts, this Chinese store offers an excellent selection of Yixing teapots and diminutive bowls.

Japonesque
824 Montgomery Street
San Francisco, CA 94133
(415) 391-8860

A modern Japanese art gallery that also carries some tea utensils.

Kelley and Ping
127 Greene Street
New York, NY 10012
(212) 228-1212

This Asian restaurant at the end of Greene Street in the Soho district is devoted to the sale and service of tea. Teas include Ti Kuan Yin, Gen Mai Cha, Gunpowder, and Chrysanthemum. Clay teapots and bowls are on display, and tea may be served with mooncakes from Taiwan, custard and coconut tarts, and an assortment of cookies.

Kobo
812 E Roy
Seattle, WA 98102
(206) 726-0704
This delightful shop sells a variety of traditional tea utensils from Japan, including tea bowls, caddies, whisks, tea cloths, and good powdered green tea. It is one of the few shops I have found in the United States that imports quality Japanese teas (*matcha*, *sencha*, *hojicha* and *genmaicha*) and a seasonal selection of dry sweets (*higashi*). The tea is sold by the ounce, packaged in a tin, and readily shipped to out-of-state customers.

Matcha and More, Inc.
4901 W. Warwick Ave.
Chicago, IL 60641
Toll Free: 877-534-0505
www.MATCHAandMORE.com
A variety of good powdered tea (*matcha*) is available as well as utensils for *Chado,* accessories, and books on *Chado*. Visit their website for more information.

Mikado
Japan Center
1737 Post Street
San Francisco, CA 94115
(415) 9229450
Offers some basic tea utensils for practice.

Mikawaya
Yaohan Plaza
333 S Alameda Street
Los Angeles, CA 90013
(213) 613-0611
Specializes in traditional Japanese tea sweets.

Minamoto Kitchoan
Swiss Center Building
608 Fifth Avenue
New York, NY 10020
(212) 489-3747
www.kitchoan.com
See below.

Minamoto Kitchoan
Mitsuwa Market Place
595 River Road
Edgewater, NJ 07020
(201) 313-9335
Offering a variety of Japanese sweets from traditional culinary arts, Minamoto Kitchoan's desserts also reflect new ideas resulting in a creative series of fruit-based, eye-catching, beautiful seasonal confectioneries.

Sara Inc.
952 Lexington Avenue
New York, NY 10021
(212) 772-3243
This beautifully designed store reflects Japanese aesthetics, and sells modern teaware from Japan as well as trays and dishes for meals. The store attracts local potters and those interested in the ceremonial service of tea, both Japanese and English-style.

Shirokiya
Ala Moana Shopping Center
Honolulu, HI 96826
(808) 973-9111
   See below.

Shirokiya
Kaahumanu Center
Kahalui, Maui, HI 96732
(808) 877-5551

Similar to Yaohan in Los Angeles, this store features a variety of Japanese goods, including kitchenware, food, and some traditional tea items. Green tea is available, although the quality of the powdered tea is questionable.

Sur La Table
84 Pine Street
Seattle, WA 98101
(206) 448-2244

Mainly sellers of fine kitchenware and equipment, the store carries teapots, other tea utensils, and tea bricks—pressed tea leaves reminiscent of the way tea was packaged in ancient China for trade along the silk route. These patterned bricks are three-quarters of an inch thick and contain two and a half pounds of concentrated tea. A catalog is available upon request.

The Tao of Tea
3430 SE Belmont St.
Portland, OR
(503) 736-0119
  See below.

The Tao of Tea
239 NW Everett St.
Portland Classical Chinese Garden
Portland, OR 97209
(503) 224-8455
  See below.

The Tao of Tea
2112 NW Hoyt St.
Portland, OR
(503) 223-3563

Offering a new approach to the purchase and enjoyment of tea, the Tao of Tea features its own tea cuisine along with an amazing variety of teas in its three unusual locations.

The Tea Box at Takashimaya
693 Fifth Avenue
New York, NY 10022
(212) 350-1500
Located on the lower level of the Japanese department store are two elegant tea rooms with an atmosphere of serenity and repose. Afternoon tea is available, and teaware, lacquerware, and linens are on sale. Behind the counter are prized loose teas, which may be purchased by the ounce.

Tea Circle
8657 Lancaster Dr.
Rohnert Park, CA 94928
(707) 792-1946
www.tea-circle.com
An online resource featuring a full line of Japanese tea ceremony items including utensils and tea from Japan as well as fine ceramics made by Northern California artists.

Tea Traders
515 SW Broadway
Portland, OR 97205
(503) 220-8533
This tea place offers a marvelous selection of choice teas and tea accessories, including Yixing teapots.

The Tea Zone
510 NW 11th Ave.
Portland, OR 97209
(503) 221-213
A tea shop selling and serving a rich variety of teas from China and Japan.

Ten Ren Tea Company
727 N Broadway #136
Los Angeles, CA 90012
(213) 626-8844
Stocks a variety of fine Chinese and Taiwanese teas and some utensils.
Brochures available.

Ten Ren Tea Company
949 Grant Avenue
San Francisco, CA
(415) 362-0656 or (800) 543-2885
This tea shop in the heart of San Francisco's Chinatown offers tea tasting, and you may experience a *gong fu* tea with one of the fine selections of oolong on sale. Utensils for the *gong fu* ceremony, as well as a variety of other Chinese and Taiwanese teas, are also available.

Ten Ren Tea and Ginseng Company
2247 S Wentworth Avenue
Chicago, IL 60616
(312) 842-1171
   See below.

Ten Ren Tea and Ginseng Company
75 Mott Street
New York, N.Y.
Fine teas from China and Taiwan are available. Here, tea tasting is offered at the rear of the store, and free courses in tea tasting are given. Contemporary teaware is on display. You may also buy Taiwanese tea sets, which include tiny teapots.

Toraya
17 E 71st Street
New York, NY 10021
(212) 861-1700
This tea shop sells Japanese confections, including the sweets for a tra-

ditional tea gathering where whisked green tea is served. Immaculately designed, the Toraya also has a tea room at the rear, its atmosphere enhanced by a skylight and tall windows. Here you can enjoy delicate green tea with some of Toraya's enticing sweets.

Upton Tea Imports
231 South Street
Hopkinton, MA 01748
(800) 234-8327
Sellers of teas from India, China, Sri Lanka (Ceylon), Formosa, and Japan, as well as teaware, including Chatsford Bone China teapots and accessories. Their catalog fully describes the fine variety of teas they carry.

Uwajimaya
600 5th Ave. South, Suite 100
Seattle, WA 98104
(206) 624-6248
  See below.

Uwajimaya
10500 SW Beaverton-Hillsdale Hwy.
Beaverton, OR 97005
(503) 643-4512
www.uwajimaya.com
Uwajimaya stores offer a variety of Asian groceries, goods and gifts including teas and utensils from Japan. It also has a large bookstore specializing in Asian language books and periodicals.

Yaohan
Little Tokyo Store
Yaohan Plaza
333 S Alameda Street
Los Angeles, CA 90013
(213) 613-0611

This large department store features a variety of Japanese kitchenware and some traditional tea utensils. Some powdered green tea is available.

For Matcha in Kyoto, Japan

Gion Tsujiri
Shijo dori, Gion machi, minamigawa
Higashi-ku, Kyoto, Japan
(075) 529-1122 or fax: (075) 529-1223

Gosho-No-En
Marutamachi, Karasuma, nishi-iru
Nakagyo-ku, Kyoto, Japan
(075) 321-1777 or fax: (075) 256-9770

Koyama-En
86 Terauchi, Ogura-cho
Uji-city, Kyoto, Japan 611
(077) 421-3151 or fax: (077) 428-2288

For the Study of Chado
The Urasenke Foundation has worldwide branches listed below:

Urasenke Konnichian
Headquarters
Ogawa Teranouchi Dori
agaru, Kamikyo-ku
Kyoto, 602 Japan
(075) 431-3111

Urasenke Hawaii Branch
245 Saratoga Road
Honolulu, HI 96815
(808) 923-1057

Urasenke Chanoyu Center
153 E 69th Street
New York, NY 10021
(212) 988-6161

Urasenke Foundation of California
2143 Powell Street
San Francisco, CA 94133
415-433-6553

Urasenke Seattle Branch
1910 Thirty-seventh Place East
Seattle, WA 98112
206-324-1483

Urasenke Foundation Vancouver Branch
209 Jackson St.
Coquitlan, BC. V3K 4C1
(604 224-1560

Urasenke Washington, D.C.
6930 Hector Road
McLean, VA 22101
(703) 748-1685

Urasenke/Mexico City
fuego 691
Col. Jardines del Pedregal
01900 Mexico D.F.
(5) 5652-5725

For an Experience of Chado

Green Gulch Farm Zen Center
1601 Shoreline Highway
Sausalito, CA 94965
415-383-3134

Japanese Garden Society of Oregon, Inc.
P.O. Box 3847
Portland, OR 97208
503-223-4070

Japanese Garden
Washington Park Arboretum
Lake Washington Boulevard East
Seattle, WA
206-684-4725

Seattle Art Museum
100 University Street
Seattle, WA 98101
206-654-3180

Wakai Association
1633 SW Skyline Blvd.
Portland, OR 97221
(503 228-7957

Acknowledgments

It is only right that I pay homage to the many writers who through the ages have recorded the evolution of tea and how the tea ceremony has been adapted and refined. In several important ways, this work is based on their efforts. It was first inspired by *The Classic of Tea*, written in China in the eighth century B.C. by Lu Yu. His masterpiece eventually led me to *The Book of Tea*, by Okakura Kakuzo, and countless other works.

I could not have written this book without the help and support of Susan Applegate, Ted Bagley, Philippa Brunsman, Evelyn Hicks, Rinko Ikeda Jeffers, Mark Ong, and Dorothy Wall. Thanks also to Christy Bartlett, Hillary Briggs, Sandor Burstein, David Charlsen, John Chen, Suzanne Copenhagen, Katherine L. Kaiser, Candice McElroy, the Missouri Historical Society in St. Louis, Bonnie Mitchell, Linda Nelson, Terence O'Donnell, Floyd Ottoson, Alan Palmer, Jennifer Reavis, Gabriele Sperling, Peyton Stafford, Laurell Swails, Nancy Trotic, Jan Waldmann, and Randy Weingarten. I wish to thank the staff at the Urasenke Foundation in Kyoto who helped with my study and research. Above all, I am also indebted to Nakamura Sensei and all my teachers of *Chado*, especially Yumiko Ban, for their kind patience and wise attention; and to my late friend Jerry Fuller, who taught me much along the Way.

Special thanks to Evelyn Hicks for her perceptive illustrations and faithful support of this book, and to PKB Associates, the *skysociety*™, *Borderline* magazine, and the McNeill Family Trust for their generous aid in this project.

Bibliography

Anderson, Jennifer. *An Introduction to Japanese Tea Ritual.* New York: State University of New York Press, 1991.

"Antioxidants in a Cup of Tea?" *University of California at Berkeley Wellness Letter* (January 1992).

Ashton, Dore. *Noguchi East and West.* Berkeley: University of California Press, 1992.

Bancroft, Anne. *Zen, Direct Pointing to Reality.* New York: Thames and Hudson, 1980.

Barnett, Robert. "A Proper Cup." *American Health* (January-February 1990).

Basho, Matsuo. *The Narrow Road to the Deep North,* translated by Nobuyuki Yasa. New York: Viking Penguin Inc., 1966.

Blofeld, John. *The Chinese Art of Tea.* Boston: Shambhala, 1985.

The Book of Tea. Paris: Flammarion, 1992.

Campbell, Joseph. *The Hero with a Thousand Faces.* Princeton, N.J.: Princeton University Press, 1973.

Castile, Rand. *The Way of Tea.* New York: John Weatherhill, 1971.

Chow, Kit, and Ione Kramer. *All the Tea in China.* San Francisco: China Books, 1990.

Clark, Garth. *The Eccentric Teapot.* New York: Cross River Press, 1989.

Crellin, John K., and Jane Philpott. *Herbal Medicine Past and Present.* Vol. I. Durham, N.C.: Duke University Press, 1990.

Cunningham, Nancy Brady. *Feeding the Spirit.* San Jose, Calif.: Resource Publications, 1988.

Deng, Ming-Dao. *ZEN: The Art of Modern Eastern Cooking.* London: Pavilion Books, 1998.

_____. *Scholar Warrior: An Introduction to the Tao in Everyday Life.* New York: HarperCollins Publishers, 1990.

Fletcher, David Wilson. *Himalayan Tea Garden.* New York: Thomas Y. Crowell Company, 1955.

Goodwin, Jason. *A Time for Tea: Travel through India and China in Search of Tea.* New York: Alfred A. Knopf, 1991.

Gustafson, Helen. *The Agony of the Leaves*. New York: Henry Holt and Company, 1996.

Hamill, Sam. *Basho's Ghost*. Seattle, WA: Broken Moon Press, 1989.

Hammitzsch, Horst. *Zen in the Art of the Tea Ceremony*. New York: St. Martin's Press, 1980.

Hynes, Angela. *The Pleasures of Afternoon Tea*. Los Angeles: HP Books, 1987.

Kapleau, Philip. *The Three Pillars of Zen*. New York: Anchor Books, 1980.

King, M. Dalton. *Special Teas*. Philadelphia: Running Press, 1992.

Lam, Kam Chuen, Master, with Lam Kai Sin and Lam Tin Yu. *The Way of Tea: The Sublime Art of Oriental Tea Drinking*. New York: Barron's Educational Series, Inc., 2002.

McCormick, Malachi. *A Decent Cup of Tea*. New York: Clarkson N. Potter, 1991.

Murai, Yasuhiko. "A Biography of Sen Rikyu." *Chanoyu Quarterly* 61 (1990).

Niles, Bo, and Veronica McNiff. *The New York Book of Tea*. New York: City and Company, 1995.

Northcote, Lady Rosalind. *The Book of Herb Lore*. New York: Dover Publications, 1971.

Okakura, Kakuzo. *The Book Of Tea*. Charles E. Tuttle Co., Inc., 1956.

Pelletier, Kenneth R. *Holistic Medicine: From Stress to Optimum Health*. New York: Delacorte Press, 1979.

_____. *Mind as Healer, Mind as Slayer: A Holistic Approach to Preventing Stress Disorders*. New York: Delacorte Press, 1977.

Perry, Sara. *The Tea Book*. San Francisco: Chronicle Books, 1993.

Plaidy, Jean. *It Began in Vauxhall Gardens*. New York: G.P. Putnam's Sons, 1955.

Pratt, James Norwood. *The Tea Lover's Treasury*. San Francisco: 101 Productions, 1982.

Riccardi, Victoria Abbott. *Untangling My Chopsticks*. New York: Broadway Books, 2003.

Rubin, Ron, and Avery Stuart. *Tea-Chings. The Tea and Her Companion: Appreciating the Varieties and Virtues of Fine Tea and Herbs*. New York, New York: Newmarket Press, 2002.

Sadler, A. L. *Cha-No-Yu: The Japanese Tea Ceremony*. Rutland, Vt.: Charles E. Tuttle Company, 1962.

Saekel, Karola. "Courting Tea Connoisseurs." *San Francisco Chronicle*, December 1, 1993.

Sen Shoshitsu XV. *The Spirit of Tea*. Tankosha, Japan, 2002.

_____. *The Japanese Way of Tea: From Its Origins in China to Sen Rikyu*. Honolulu, HI: University of Hawaii Press, 1998.

_____. "The Diffusion of the Way of Tea." *Chanoyu Quarterly* 73 (1993).

_____. *Chanoyu: The Urasenke Tradition of Tea*. New York: John Weatherhill, 1988.

_____. *Chado: the Japanese Way of Tea*. New York: Weatherhill/Tankosha, 1979.

_____. *Tea Life, Tea Mind*. New York: John Weatherhill, 1979.

Sen'o Tanaka. *The Tea Ceremony*. Tokyo: Kodansha International, 1973.

Shalleck, Jamie. *Tea*. New York: Viking Press, 1971.

Shigenori, Chikamatsu. *Stories from a Tea Room*. Rutland, Vt.: Charles E. Tuttle Company, 1982.

Simpson, Helen. *The London Ritz Book of Afternoon Tea: The Art and Pleasures of Taking Tea*. New York: Arbor House, 1986.

Smith, Jonathan C. *Spiritual Living for a Skeptical Age: A Psychological Approach to Meditative Practice*. New York: Insight Books, 1992.

Smith, Michael. *The Afternoon Teabook*. New York: Macmillan, 1986.

Snider, Sharon. "Beware of the Unknown Brew: Herbal Teas and Toxicity." *FDA Consumer* (May 1991).

Sogyal, Rinpoche. *The Tibetan Book of Living and Dying*. New York: HarperCollins Publishers, 1992.

Suzuki, D. T. *Essays in Zen Buddhism*. New York: Samuel Weiser, 1970.

_____. *Living by Zen*. New York: Samuel Weiser, 1950.

Tanizaki, Jun'ichiro. *In Praise of Shadows*. Rutland, Vt.: Charles E. Tuttle Company, 1977.

Ukers, William H. *The Romance of Tea*. New York: Alfred A. Knopf, 1936.

Varley, Paul, and Kamakura Isao, eds. *Tea in Japan: Essays on the History of Chanoyu*. Honolulu: University of Hawaii Press, 1989.

Vitell, Bettina. *The World in a Bowl of Tea*. New York: HarperCollins Publishers, 1997.

Watts, Alan. *Haiku and Zen/Senryu*. Audiotape. Sausalito, Calif.: MEA Catalog Recordings.

"*Which Tea Is Best?*" Consumer Reports (July 1992).

Wilson, Anne C. *Food and Drink in Britain: From the Stone Age to the Nineteenth Century*. London: Constable and Company, 1973.

Wong, Eva. *The Shambhala Guide to Taoism*. Boston: Shambhala Publications, 1997.

Woodward, Nancy Hyden. *Teas of the World*. New York: Collier Books, 1980.

Wu, K. C. *The Chinese Heritage*. New York: Crown Publishers, 1982.

Yu, Lu. *The Classic of Tea*. Translated by Francis Ross Carpenter. Boston: Little, Brown and Co., 1974.

Ziegler, Mel. *The Republic of Tea: Letters to a Young Entrepreneur*. New York: Doubleday/Currency, 1992.